Practical Analog Electronics for Technicians

Practical Analog Electronics for Technicians

Will Kimber

Routledge
Taylor & Francis Group

LONDON AND NEW YORK

First published by Butterworth-Heinemann

First published 1997

This edition published 2011 by Routledge
2 Park Square, Milton Park, Abingdon, Oxon OX14 4RN
711 Third Avenue, New York, NY 10017, USA

Routledge is an imprint of the Taylor & Francis Group, an informa business

British Library Cataloguing in Publication Data
Kimber, Will
    Practical analog electronics for technicians
    1.   Analog electronic systems
    I.   Title
    621.3'815

ISBN 0 7506 2952 5

Typeset by Laser Words, Madras, India

Contents

Preface

This book combines an explanation of the basic theory of analog techniques with a total of 50 practical exercises, all of which can be carried out with equipment that is normally readily available in departments running electronic courses.

These exercises are intended to provide the opportunity to become familiar with a wide range of analog circuits and are based on my own experience of teaching electronics at all levels of further education. I hope that readers will forgive the 'cookbook-type' instructions.

Throughout the book there are many questions (answers provided, where appropriate); some of which require further research and reading. A knowledge of electrical principles to BTEC level II, or its equivalent, is assumed.

The book is intended to cater for students following Advanced GNVQ in Electronics, BTEC NII/III, GCE A-level and City & Guilds courses.

Will Kimber

Practical exercises

1 | Setting the scene

Introduction

For those electrical engineers who did their initial training before digital electronics made such an impact, the analog way **was** the way of electrical life. Quite simply, that's all there was for a while. Digital electronics, although not unknown, did not become an economic proposition until the advent of the integrated circuit in the 1960s. A question asking 'what is analog all about?' hardly seemed relevant.

Let us ask the question now. Just what **is** analog electronics and what place does it have in today's world? A dictionary definition is perhaps only moderately helpful. We see for example, that *The Concise Oxford Dictionary* gives

analog (or analogue) as 'a parallel word or thing'

analogous as 'similar or parallel to'

analogy as 'similarity'.

The more important aspect from our point of view, is the understanding of what an analog **signal** is and how it differs from the digital signal.

Digital and analog signals

The digital process recognizes a limited number of states or levels. The **two-state** logic system for example, is represented variously, by

on and **off, 1** and **0, high** and **low**

An example of a digital signal is shown in Figure 1.1.

The analog process, on the other hand, recognizes a continuously variable state, where the levels are, theoretically at least, infinitely variable between a lower and an upper limit. An example of an analog signal is shown in Figure 1.2.

A recognizable example is the speedometer in a car, in which the pointer can take **any position** between zero miles per hour and the maximum possible (or maximum allowed). The analog or **analogy** here, is the position of the pointer representing the quantity, speed. And of more immediate relevance, there is the moving coil meter, the pointer of which moves to a position related to the current through the coil. It is the presence of an **indicator**, such as the pointer, that helps the identification of an analog system.

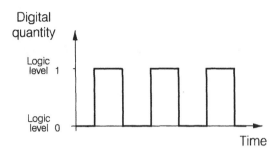

Figure 1.1 *The digital signal*

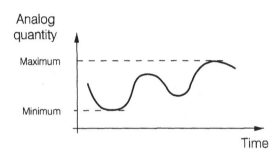

Figure 1.2 *The analog signal*

As far as the place for analog electronics in today's world is concerned, it can be mentioned that most of the natural things we measure, such as temperature, wind speed, the human voice, vibrations (or oscillations, for which see Chapter 7), are all analog systems.

Referring again to pre-integrated circuit days, the only way of processing these systems was with analog circuits, of which amplification was one example.

With today's technology it is frequently the case of converting an analog signal into a digital format and converting back again where necessary. There are many advantages in this technique which cannot be dealt with here.

The practical exercises

Typical equipment and the components needed are listed for each practical exercise.

The power supply voltages chosen for these exercises are those that, in the absence of a commercial unit, can be made up from easily available fixed and variable voltage regulators. The details for this are readily available in a variety of texts.

It will be seen that a dual-polarity (±V) supply is needed for the operational amplifier circuits. The requirement here is to provide both a positive and a negative voltage with respect to the 0 V line. This can, if necessary, be achieved with two single-polarity supplies as shown in Figure 1.3.

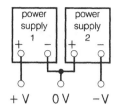

Figure 1.3   *A dual power supply from two single power supplies*

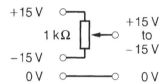

Figure 1.4   *A single power supply from a dual power supply*

For the exercises that need a supply voltage, variable from +V through zero to −V, this can be obtained by the method shown in Figure 1.4.

Waveform observation is always important, not least in fault finding, where even the absence of a signal can be revealing. A **cathode ray oscilloscope** (CRO) is essential, and preferably double beam, especially when phase shift details are needed. For the majority of the exercises the CRO has not been shown on the circuit diagram, but its general purpose is to display both the input and the output waveforms.

The audio frequency **signal generator** will need to provide a sinusoidal input signal to the unit, at modest voltage levels, up to a frequency of approximately 50 kHz. Most commercial units will do this in spite of this frequency being well outside the accepted audio frequency range. In addition, a square waveform at 10 V is required for one practical exercise.

The actual **meter** used for DC voltage and current measurements can be a personal choice of the user for the majority of the practical exercises. It would be more helpful to use the digital type wherever the need to avoid circuit loading is paramount (see Chapter 8). For fault finding in particular, the use of a digital voltmeter, when attempting to prove the existence of a short circuit, can overcome any difficulty in determining the precise reading on an analog meter. For a brief comparison between digital and analog meters, the reader is referred again to Chapter 8.

The manner of assembly of a particular circuit is, of course, a matter of choice for the practitioner. It is believed that the learning experience **can** be enhanced by a certain amount of self-assembly within the constraints of the time available. With this in mind, there are breadboard systems available where component wires push into holes on a 0.1 inch grid. Although this system is mostly convenient, it can, on occasions, also be the source of much frustration and despair!

A final word concerning the 'batting order' for the chapters. The natural order can be followed with safety, since it provides a natural progression of knowledge obtained through the chapters. The exception would be Chapter 8 – Test and measuring equipment – which can be taken at any stage, even first, if desired.

2 DC power supply units

Although the use of thermionic valves may be making a comeback in some quarters, the great attraction of transistors and integrated circuits in, for example, personal stereos and other low-output power equipment, is the 'carry everywhere' aspect, which is in no small measure due to the low-voltage, low-power supply that the equipment requires.

As most users will know, this power supply can easily be provided by batteries, with the cost aspect considered by the use of rechargeable batteries. The wisdom of using a mains adaptor to reduce running costs to a minimum is advocated, although of course, portability is reduced.

For higher output powers, as offered by domestic high fidelity equipment, the use of batteries is not an economic proposition. The need therefore is for the **DC power supply unit** to be mains-derived, which is the subject of this chapter.

Questions and answers

What is a DC power supply?

A DC (direct current) power supply is an item of equipment which provides a unidirectional (in one direction only) voltage which ideally should be constant.

Why is a DC power supply necessary?

Electronic equipment requires an operating supply voltage which is of the type described above. This voltage could range in value from a few hundred volts in the days of thermionic valves, to between three and, say, 50 volts for modern-day equipment.

This supply voltage would not normally be provided sensibly and economically with batteries.

What is the function of a DC power supply?

To provide a constant (or near constant) voltage while delivering a possible wide range of currents to the equipment. The actual current demand can **sometimes** depend upon the operating state of the equipment – for example, with an audio amplifier, the actual volume level.

It can be mentioned here that there are certain applications which require a constant current supply as opposed to constant voltage.

How is this DC power supply achieved?

From the AC (alternating current) mains, through a conversion process known as **rectification**, illustrated in Figure 2.1.

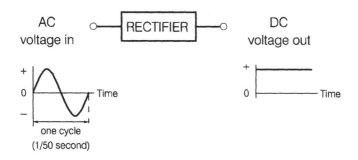

Figure 2.1 *The rectifier block diagram*

Referring to the diagram in Figure 2.1:

The AC input voltage is shown as alternating **above and below** zero, that is, going alternately positive and negative, at a rate (frequency) of 50 times a second (50 hertz).

The DC output voltage is shown as a positive value, but it could, if necessary, be negative.

Rectification

The essential item in the rectification process is the **diode**.

The word 'diode' is made up from **di** (= two) + electr**ode**, and it is therefore a device which has two electrodes or terminals. These terminals are called anode and cathode respectively and the circuit symbol is shown in Figure 2.2(a). Identification of the cathode is by means of marking the appropriate end of the body of the component, usually with a band, as shown in Figure 2.2(b).

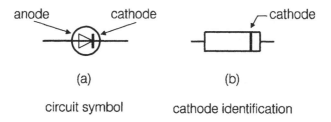

Figure 2.2 *The diode: circuit symbol and cathode identification*

The diode is a **one-way** device, that is, it will conduct, or allow current to flow, in one direction only, when its anode is **positive** with respect to its cathode. The direction in which this current flows is called the **forward** direction.

When the anode is **negative** with respect to the cathode, no current should flow. This is called the **reverse** direction.

The diode can be regarded as a switch, with current able to flow only when the switch is on.

Important points

- In theory, the ideal diode (or switch) will have:
  (a) zero resistance in the forward (conducting) direction
  (b) infinite resistance in the reverse (non-conducting) direction.

- In practice, the diode will have:
  (a) a small resistance, typically several ohms, in the forward direction. It is therefore not a perfect switch.
  (b) less than infinite resistance in the non-conducting direction, typically several megohms. As a result, there will be a small current flow in the reverse direction after all. This current is known as **leakage**.

- The direction of the 'arrow head' in the circuit symbol (Figure 2.2a), shows the direction of forward current flow.

- Semiconductor diodes are made from either germanium or silicon. They are referred to as **pn junction** diodes, following the physics of their manufacture.

- Rectifying diodes are classified as being either signal or power types. While signal diodes can be either germanium or silicon, it is usual for power diodes to be silicon.

Practical Exercise 2.1

To obtain the forward and reverse characteristics of signal and power diodes.
  For this exercise you will need the following components and equipment:

1 – silicon diode (1N4001)
1 – germanium diode (0A91)
1 – DC power supply (variable from 0 to +25 V)
1 – resistor (10 Ω, 3 W)
1 – variable resistor (1 kΩ)
1 – DC ammeter
1 – DC voltmeter

Procedure (a) Forward characteristic of the silicon diode

1 Connect up the circuit shown in Figure. 2.3(a) making sure that the variable resistor RV_1 is set to its maximum value.
2 By reducing the value of RV_1, obtain a series of current (I_f) and voltage (V_f) readings and complete the table shown in Figure 2.3(b).

Procedure (b) Reverse characteristic of the silicon diode

1 Connect up the circuit shown in Figure 2.4(a).

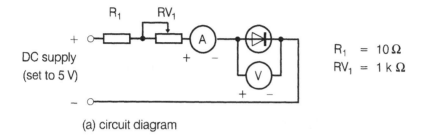

(a) circuit diagram

R_1 = 10 Ω
RV_1 = 1 k Ω

Current through diode (I_f)									mA
Voltage across diode (V_f)									V

(b) table of results

Figure 2.3 *Forward characteristic of the diode: circuit diagram and table of results for Practical Exercise 2.1(a)*

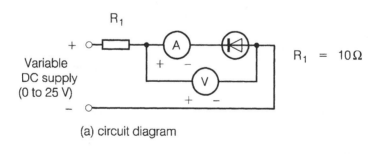

(a) circuit diagram

R_1 = 10 Ω

Current through diode (I_r)							μA
Voltage across diode (V_r)							V

(b) table of results

Figure 2.4 *Reverse characteristic of the diode: circuit diagram and table of results for Practical Exercise 2.1(b)*

Continued on p. 8

Practical Exercise 2.1 *(Continued)*

2 Increase the supply voltage in steps from zero to a maximum of 25 V and enter the corresponding values of current (I_r) and voltage (V_r) in the table in Figure 2.4(b).

Procedures (c) and (d)

1 Repeat procedures (a) and (b) for the germanium diode but **do not** exceed a forward current (I_f) of 50 mA.

Procedure (e) Drawing the graph

A typical graph for each diode is shown in Figure. 2.5. Note the use of different scales for the current and voltage axes.

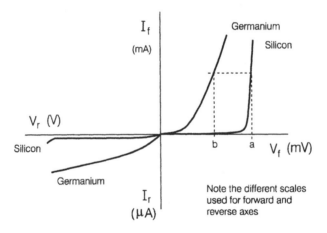

Figure 2.5   *Typical characteristic for the silicon and germanium diode*

Conclusions

1 From your graph, estimate the value of the forward voltage (V_f) corresponding to points A and B respectively on the graph in Figure 2.5.
2 Compare the values of the reverse currents of each diode at a reverse voltage of 25 V. Which diode would you suggest has the higher resistance to reverse current flow?
3 What do you think would happen to the diodes if either the forward current or the reverse voltage was increased well beyond the limits shown in their respective tables?
4 What is the reason for making the voltmeter connection in Figure 2.4(a) different from that in Figure 2.3a?

Important points

- The ability of a diode to conduct in the forward direction is shown by its **forward current** rating.

- The forward voltage drop depends upon the forward current. For germanium, the variation in voltage drop is typically from 0.1 V to 0.4 V, while for silicon it is from 0.4 V to 0.7 V.

- The ability of a diode to withstand reverse voltages without breaking down is shown by its **reverse voltage** rating. Other names are used for this, such as peak inverse voltage (PIV) and reverse repetitive voltage (V_{RRM}) and will be referred to again later.

- The requirement for a signal diode is a low forward voltage drop, with its use being for low-level detection in radio frequency and signal processing applications.

- The requirements for a power diode are high forward current and high reverse voltage ratings for its application as a rectifier in DC power supplies.

Question

2.1 Obtain the manufacturer's data sheets or component supplier's catalogues and complete the table given in Figure 2.6.

Type	Material	Max I_f	PIV (V_{RRM})	Max I_r	Application
OA90					
1N4148					
1N4004					
1N4006					
1N5401					
WO1					

Figure 2.6 *Semiconductor diodes: table for data for Question 2.1*

Practical Exercise 2.2

To investigate the action of the diode as a one-way device.

Continued on p. 10

Practical Exercise 2.2 *(Continued)*

For this exercise you will need the following components and equipment:

1 – silicon diode (1N4001)
1 – DC supply (+5 V)
1 – filament lamp (6 V)
1 – DC voltmeter

Procedure (a)

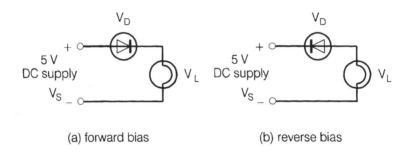

(a) forward bias   (b) reverse bias

Figure 2.7 *The diode as a one-way device: circuit diagrams for Practical Exercise 2.2*

Using the circuit shown in Figure 2.7(a), the diode is **forward biased** and the lamp should be lit.

1   Measure the supply voltage $V_S =$ _____ V
2   Measure the voltage across the lamp $V_L =$ _____ V
3   Measure the voltage across the diode $V_D =$ _____ V
                             $V_L + V_D =$ _____ V

Using the circuit shown in Figure 2.7(b), the diode is **reversed biased** and the lamp should **not** be lit!

4   Measure the supply voltage $V_S =$ _____ V
5   Measure the voltage across the lamp $V_L =$ _____ V
6   Measure the voltage across the diode $V_D =$ _____ V
                             $V_L + V_D =$ _____ V

Important points

- The diode can be regarded as a switch, which is **closed** when the anode is positive with respect to the cathode (Figure 2.7(a)) and **open** when the anode is negative with respect to the cathode (Figure 2.7(b)).

- In the forward biased, or conducting state, the forward voltage drop across the (silicon) diode should be approximately 0.6 V, with almost all of the supply voltage being across the lamp. The sum of V_L and V_D should equal the supply voltage V_S.

- If a germanium diode had been used, this forward voltage would have been approximately 0.2 V.

- The forward current (I_L) will be given by

$$I_L = \frac{V_S}{R_L + R_f}$$

where R_f represents the diode forward resistance and R_L the resistance of the (lamp) load.

- In the reverse biased, or non-conducting state, the voltage across the lamp should be zero, with all of the supply voltage being across the diode.
  The sum of V_L and V_D should equal the supply voltage V_S.

- The reverse (leakage) current may, in practice, not be zero and as a result there would be a small voltage across the lamp.

- The situation *'the anode is positive with respect to the cathode'* is identical to *'the cathode is negative with respect to the anode'*.

Procedure (b)

1 Measure the forward and reverse resistance of each diode.
   Note: If a moving-coil meter is used, remember to pay particular attention to the polarity of the internal battery.

silicon diode: forward resistance = _____ Ω

                                     reverse resistance = _____ Ω

germanium diode: forward resistance = _____ Ω

                                     reverse resistance = _____ Ω

Questions

2.2 Refer to Figure 2.8(a)–(c) inclusive. State which lamp(s) will be lit.
2.3 Refer to Figure 2.8(d). State which diode(s) will be conducting.
2.4 Refer to Figure 2.8(e)–(g) inclusive. Complete the table.

Continued on p. 12

Questions *(Continued)*

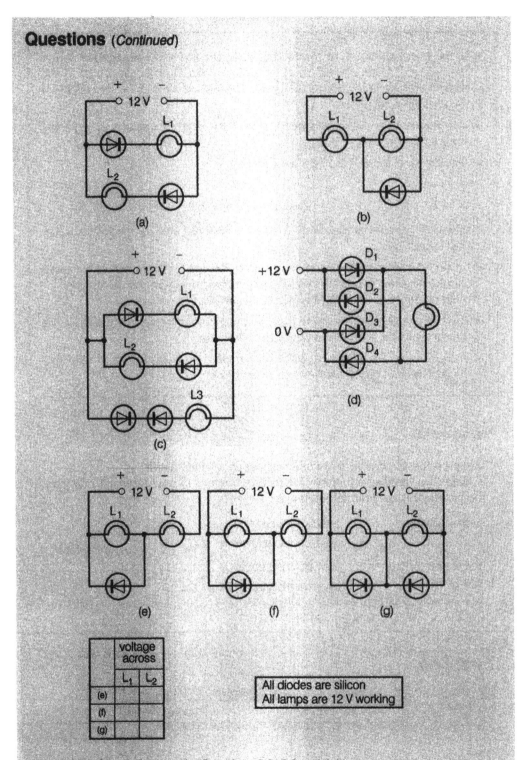

Figure 2.8 *Circuit diagrams for Questions 2.2, 2.3 and 2.4*

The following practical exercise will show the use of the diode in converting an AC supply to a DC supply.

Practical Exercise 2.3

To investigate half-wave and full-wave rectification.
    For this exercise you will need the following components and equipment:

1 – transformer unit (240 V primary, 12 V secondary with centre-tap)
4 – silicon diode (1N4001)
1 – filament lamp (6 V)
1 – capacitor (100 µF, 220 µF, 1000 µF, 2200 µF, all 16 V working)
1 – resistor (10 Ω)
1 – double beam cathode ray oscilloscope
1 – dc voltmeter

Procedure (a) Half-wave rectification

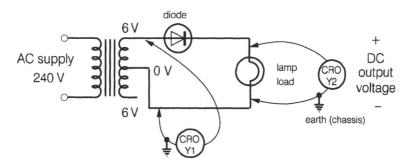

Figure 2.9 *The half-wave rectifier: circuit diagram for Practical Exercise 2.3(a)*

Make up the circuit shown in Figure. 2.9.
    Make sure that both chassis connections for the CRO go to the centre-tap of the transformer.
1 Sketch, on squared paper, the waveforms of:

(a) the secondary AC input voltage to the diode.
(b) the DC output voltage across the lamp load.

Typical waveforms for these are shown in Figure 2.10.
2 Measure with the cathode ray oscilloscope the peak level of each waveform in procedure 1 above.
3 Use the voltmeter to measure the average DC voltage across the load.
4 Enter all measured values on the appropriate waveform diagrams.

Continued on p. 14

Practical Exercise 2.3 *(Continued)*

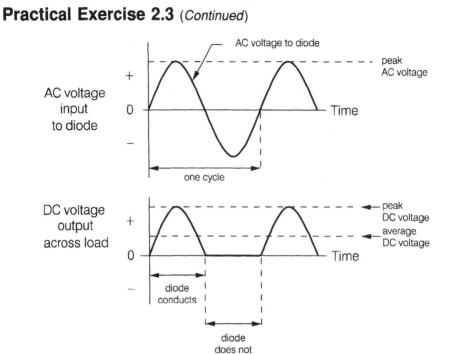

Figure 2.10   *The half-wave rectifier. No capacitor across the load: waveforms for Practical Exercise 2.3(a)*

Important point

- As it stands, the output voltage **is** DC because it is unidirectional. Unfortunately, it is not constant (steady) and as such, would be unsuitable as a DC power supply for electronic equipment. It would be satisfactory as a battery charger.

---

5 Add the 1000 μF capacitor to the circuit as shown in Figure 2.11. **Make sure** that the polarity of the capacitor is correct.
6 Note the effect on the brightness of the lamp load of adding the capacitor. Measure the DC output voltage across the load, which should be larger than the value obtained in procedure 3.
7 Sketch the waveform of the AC input voltage to the diode and the DC output voltage across the load.
  Typical waveforms are shown in Figure 2.12.
8 Estimate the peak-to-peak variation in the DC output voltage.

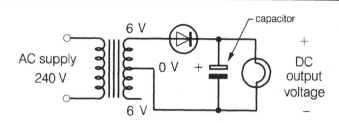

Figure 2.11 *The half-wave rectifier: Addition of capacitor for Practical Exercise 2.3(a)*

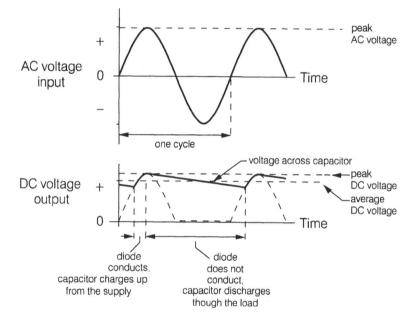

Figure 2.12 *The half-wave rectifier. Capacitor across the load: waveforms for Practical Exercise 2.3(a)*

9 Keep the circuit switched on and remove the lamp from the circuit. Note any change in the DC output voltage waveform.

Conclusion

1 For this half-wave rectifier circuit, state the value of:

    (a) the DC output voltage (with 1000 μF capacitor),

    (b) the peak AC secondary voltage and, therefore,

    (c) the rms AC secondary voltage necessary to provide this DC output voltage.

Continued on p. 16

Practical Exercise 2.3 (*Continued*)

Important points

- The DC output voltage waveform (Figure 2.12) shows the capacitor charging up through the diode and discharging through the load. The addition of the capacitor increases the DC output voltage to a value closer to that of the peak AC secondary voltage.

- The charging time, which is the time for which the diode conducts, is small compared with the discharging time.

- Removing the load means that no current is supplied from the capacitor which therefore will not need to discharge. The DC output voltage here will be equal to the peak AC secondary voltage. This is now a power supply in waiting!

- The capacitor placed in parallel with the load is known as a **reservoir** capacitor.

10 Use values of 100 µF, 220 µF, and 2200 µF in turn to investigate the effect on the DC output voltage. Complete the table shown in Figure 2.13.

Reservoir capacitor (µF)	DC output voltage (V)
None	
100	
220	
1000	
2200	

Figure 2.13 *Table of results for Practical Exercise 2.3(a)*

11 Modify the circuit to that shown in Figure. 2.14 in order to display the voltage waveform across the 10 Ω resistor. This waveform will be identical, in shape and position, to the current **through** the resistor, and will thus represent the current through the diode during the charging period. A typical waveform is shown in Figure 2.15.

12 Add this waveform to your sketch from procedure 7.

13 Measure with the CRO the time during which the capacitor charges up.

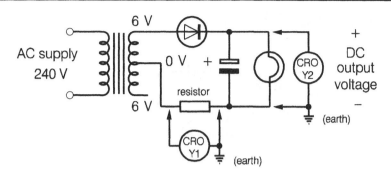

Figure 2.14 *The half-wave rectifier: to display the diode current waveform for Practical Exercise 2.3(a)*

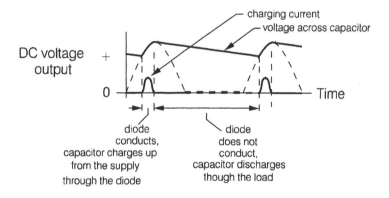

Figure 2.15 *The half-wave rectifier. Capacitor across the load: the charging current waveform*

Important points

- The variation in DC output voltage as shown in Figure 2.12 is known as **ripple**.

- For the half-wave rectifier, one complete cycle of the AC input voltage gives one charge-and-discharge cycle. The frequency of the ripple voltage is thus the same as that of the AC input voltage, namely, 50 hertz (50 Hz).

Procedure (b) Full-wave rectification (centre-tap)

Make up the circuit shown in Figure 2.16.
 1 Sketch on squared paper the waveforms of the secondary AC input voltage to **each** diode and the DC output voltage across the load.

Continued on p. 18

Practical Exercise 2.3 (*Continued*)

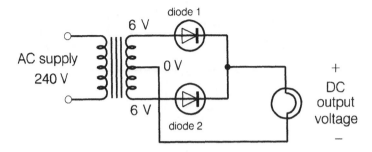

Figure 2.16   *The full-wave, centre-tap rectifier: circuit diagram for Practical Exercise 2.3(b)*

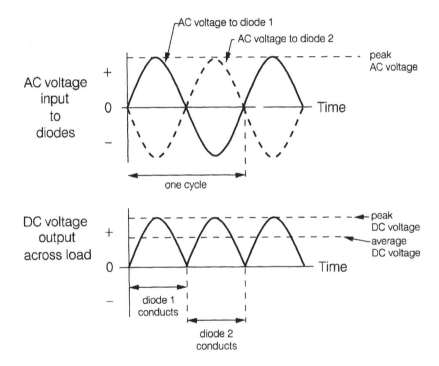

Figure 2.17   *The full-wave, centre-tap rectifier. No capacitor across the load: waveforms for Practical Exercise 2.3(b)*

Typical waveforms for these are shown in Figure 2.17.

2 Measure with the oscilloscope the peak level of each waveform in procedure 1 above.

3 Use the voltmeter to measure the DC voltage across the load.

4 Enter all measured values on the waveform diagrams.

5 Connect the 1000 μF capacitor in parallel with the load as before, making sure that the polarity is correct.

6 Note the effect on the brightness of the lamp of adding the 1000 μF capacitor. Measure the DC output voltage. (This time, any change of brightness should not be so pronounced.)

7 Sketch the waveforms of the AC input and DC output voltages. Typical waveforms are shown in Figure 2.18.

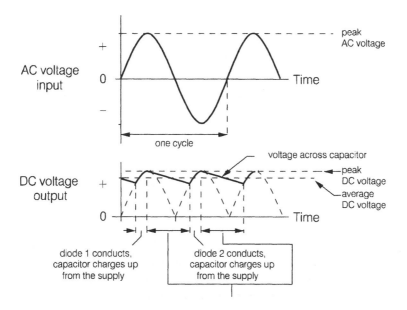

Figure 2.18 *The full-wave centre-tap rectifier. Capacitor across the load: waveforms for Practical Exercise 2.3(b)*

8 Estimate the peak-to-peak variation (ripple) in the DC output voltage.

9 Modify the circuit to that shown in Figure 2.19 in order to display the voltage waveform across the 10 Ω resistor. This waveform will represent the current through each diode separately during their respective charging periods.

10 Add this waveform to your sketch from procedure 7.

11 Measure with the CRO the charging time for each diode. This should be less than that measured previously in Procedure (a)13.

Conclusions

1 For this full-wave rectifier circuit, state the value of:

(a) the DC output voltage (with 1000 μF capacitor),
(b) the peak AC secondary voltage and, therefore,

Continued on p. 20

Practical Exercise 2.3 (*Continued*)

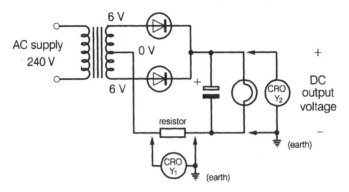

Figure 2.19 *The full-wave, centre-tap rectifier: to display the diode current waveforms for Practical Exercise 2.3(b)*

(c) the **total** rms AC secondary voltage necessary to provide this DC output voltage.

2 State the frequency of the ripple voltage of the full-wave centre-tap rectifier.

Important points

- Charging of the capacitor occurs only when the secondary AC input voltage is greater than the voltage across the capacitor. Since this charging time is small (µs), the actual charging current required to replace the lost charge can be large. This follows from the relationship:

$$\text{current } (I) = \frac{\text{charge } (Q)}{\text{time } (t)} \text{ or } I = \frac{Q}{t}$$

Thus, for a given charge (Q), the smaller the value of time (t), the larger the value of current (I).

- For the full-wave (centre-tap) rectifier, one half of the transformer secondary (i.e. 6 V) is used for each diode. The transformer secondary winding specification is written (using this example) as 6 V – 0 V – 6 V.

Question

2.5 By considering such factors as DC output voltage and ripple voltage, suggest why a full-wave rectifier centre-tap circuit might be preferred to a half-wave circuit.
Give any possible disadvantages of the full-wave type. (*Hint.* Consider the transformer secondary voltage and the number of diodes.)

Procedure (c) Full-wave rectification (bridge)

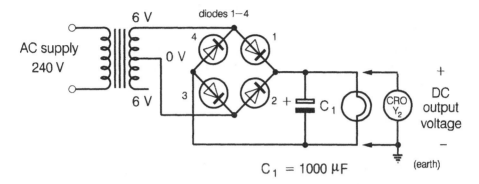

$C_1 = 1000\ \mu F$

Figure 2.20 *The full-wave bridge rectifier: circuit diagram for Practical Exercise 2.3(c)*

Make up the circuit shown in Figure. 2.20. **Do not** use the second input of the CRO to observe the AC input voltage to the diodes since there is no easy earth return for this input.

1 The waveforms of the secondary AC input voltage to each pair of diodes and the DC output voltage across the load should be similar to that obtained for the full-wave (centre-tap) circuit.
2 Use the oscilloscope and note the peak level of each AC input voltage waveform in procedure 1 above.
3 Use the voltmeter to measure and note the value of the DC voltage across the load.
4 Measure the peak-to-peak ripple voltage.

Important points

- The results from Procedure (c) should not be very different from the corresponding results from Procedure (b).

- In the bridge circuit there are two diodes conducting in each half-cycle. Referring to Figure 2.20, these diodes are D_1 with D_3 and D_2 with D_4 respectively.

- The voltage drop across a conducting silicon diode will be approximately 0.6 V.

Conclusions

1 For this full-wave rectifier circuit, state the value of:
   (a) the DC output voltage (with 1000 µF capacitor)

Continued on p. 22

Practical Exercise 2.3 (*Continued*)

(b) the peak AC secondary voltage and therefore,

(c) the rms AC secondary voltage necessary to provide this DC output voltage.

2 State the frequency of the ripple voltage of the full-wave bridge rectifier.

3 For a comparison of the three rectifier circuits, complete the table shown in Figure 2.21.

Reservoir capacitor C = 1000 µF	Total AC secondary voltage (rms)	Peak-to-peak ripple voltage	Ripple frequency	DC output voltage
Half-wave				
Full-wave centre-tap				
Full-wave bridge				

Figure 2.21 *Comparison of types of rectification: table for Practical Exercise 2.3(c)*

4 State, with reason(s), your preferred choice of circuit.

Procedure (d) The effect of increasing the load current

1 Using the circuit of Figure 2.20, investigate the effect on

(i) the DC output voltage

(ii) the peak-to-peak ripple voltage, of adding a second, then a third lamp in parallel.

2 Complete the table shown in Figure 2.22.

Important points

- The DC output voltage should **decrease** as the lamp load is increased.

- The fall in average DC output voltage creates a condition known as **regulation**.

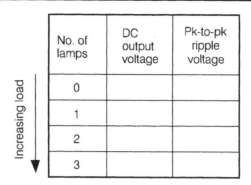

	No. of lamps	DC output voltage	Pk-to-pk ripple voltage
Increasing load	0		
	1		
	2		
	3		

Figure 2.22 *Increasing the load current: table for Practical Exercise 2.3(d)*

Reduction of ripple

An increase in value of the reservoir capacitor in Practical Exercise 2.3 Procedure (a)10 will have resulted in an increase in DC output voltage. There are practical limits, however, to the value of capacitance that can be used. A larger value of capacitance requires a larger charging current in a shorter period of time. Excessive charging current can cause permanent damage to the diodes. The inclusion of a current limiting resistor, in series with the diode, can help, but in turn, there will be a reduction in DC output voltage due to the voltage drop across this resistor, which will depend upon the actual value of load current.

Important points

- The larger the required load current, the larger the value of the reservoir capacitor must be. Typical values are given below:

Load current	Reservoir capacitance
0.5 A	2 200 µF
1.0 A	4 700 µF
2.0 A	10 000 µF

- The working voltage of the reservoir capacitor should be at least 1.4 times the no-load DC output voltage.

- The ratio of peak-to-average current can be large. For example if this ratio is 10 to 1, then the diode forward current rating must be 10 times the maximum continuous load current.

Peak Inverse Voltage (PIV) rating

Due to the presence of the reservoir capacitor, the voltage across the non-conducting diode(s), known as the **inverse** voltage, will not just be the AC negative-going half cycle.

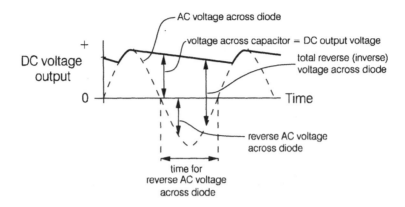

Figure 2.23   *Waveforms to illustrate reverse (inverse) voltage*

Figure 2.23 shows that the instantaneous inverse voltage will be the sum of the voltage across the capacitor and the instantaneous negative AC voltage.
   In fact,

peak-inverse-voltage (PIV)
      = maximum reverse voltage
      = the maximum voltage across the non-conducting diodes
      = the voltage across the capacitor + the maximum instantaneous AC voltage

The voltage across the capacitor will be very nearly equal to the peak AC voltage, while the maximum instantaneous AC voltage is of course equal to the peak AC voltage.

Important points

- The peak inverse voltage can be taken to be equal to **twice** the peak AC secondary voltage applied to the non-conducting diode(s).

- An alternative name for peak-inverse-voltage is **reverse repetitive voltage** (V_{RRM}).

- The diode(s) must be able to withstand this reverse repetitive voltage without breaking down.

Calculation of DC output voltage

Let us take the half-wave circuit of Figure 2.11 as an example.
   We shall assume:
(a) a typical voltage drop across the diode of 0.6 V
(b) an insignificant internal voltage drop within the transformer secondary winding
(c) the load current is small enough to allow the capacitor to remain almost fully charged, that is, at the peak value of the charging voltage (on the right-hand side of the diode).

With the values given:

transformer rms secondary voltage $= 6$ V

transformer peak secondary voltage $= \sqrt{2} \times 6$ V

$$= 8.5 \text{ V}$$

peak charging voltage $= 8.5$ V $-$ (voltage drop across the diode)

$$= (8.5 - 0.6) \text{ V}$$

$$= 7.9 \text{ V}$$

$$= \text{DC output voltage} \quad \text{(to a reasonable approximation)}$$

It would be helpful for the reader to return to Practical Exercise 2.3 (Procedures (a) to (c)) and verify the truth (or otherwise) of these assumptions.

Remember, if the capacitor does discharge substantially, then the calculations will only be approximate!

The calculations for the full-wave circuits should be more accurate since there is less time for the capacitor to discharge.

Remember:

(a) for the centre-tap circuit there is 6 V rms applied to **each** diode
(b) for the bridge circuit, there is 6 V rms applied to **two** diodes in series.

To return to the subject of PIV rating and using the above example:

(a) for the half-wave and full-wave centre-tap arrangements, the AC secondary voltage is 6 V rms, that is, 8.5 V peak. Thus the required minimum PIV rating for each diode would be 2×8.5 V, that is, 17 V.
(b) for the bridge circuit, notice there are two diodes in series sharing the reverse (inverse) voltage. Thus the minimum PIV rating for each diode would need to be one-half of that for the centre-tap circuit, that is, 8.5 V.

Calculation of transformer secondary voltage

The calculations can be applied in reverse, starting from the required DC output voltage, using the same assumptions as before, and ending with the necessary value of transformer secondary voltage.

Questions

2.6 A transformer with a 17 V – 0 V – 17 V secondary winding is used in a full-wave centre-tap rectifier circuit, with a large reservoir capacitor. Calculate the approximate value of the DC output voltage stating any assumptions made. Calculate also the minimum PIV rating for each diode.

Continued on p. 26

A typical mains-operated power supply showing essential safety features

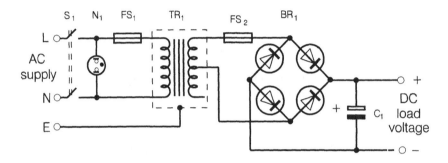

Figure 2.24   *The DC power supply: a typical circuit diagram*

This is shown in Figure 2.24. The essential items include:

(a) a transformer, to step down the 240 V incoming mains voltage to a suitable value and to isolate the mains from the rectifier circuit.
(b) a neon, with built-in resistor, to indicate 'mains-on'
(c) a double-pole switch in the supply side
(d) an appropriately rated fuse in the live conductor of the supply
(e) an earth connection to the metal case of the transformer
(f) an appropriately rated fuse in the output side.

2.11 Choose a diode from (a) to perform the task in (b).

(a)

Diode	I_f	PIV
1N5406	3 A	600 V
1N914	75 mA	100 V
BY127	1 A	1250 V
OA47	110 mA	25 V

(b) 2 A 12 V battery charger
5 V, 50 mA logic switching
DC power supply using 240 V, 0.5 A transformer secondary

Regulation

The regulation of a power supply is a measure of how the output voltage will change (decrease) as the current delivered to the load increases. Figure 2.25 shows a typical graph of load voltage against load current. Notice the steady fall in voltage as the load current increases.

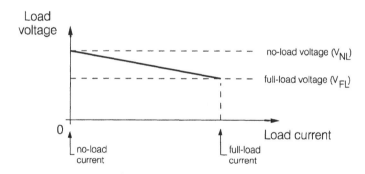

Figure 2.25 *The load characteristic of a power supply*

For a definition of regulation it is necessary to consider a change in load current from zero to its full-load (maximum) value. Thus

$$\text{regulation} = \frac{\text{change in output voltage from no load to full load}}{\text{no load output voltage}}$$

$$= \frac{V_{\text{No Load}} - V_{\text{Full Load}}}{V_{\text{No Load}}} = \frac{V_{\text{NL}} - V_{\text{FL}}}{V_{\text{NL}}}$$

Example 1

(a) The output voltage of a certain power supply decreases from 9 V to 7.5 V as the load current increases from zero to its full-load value. Calculate the percentage regulation of this power supply.

The no-load voltage (V_{NL}) is 9 V and the full-load voltage (V_{FL}) is 7.5 V. Thus the regulation is

$$\frac{V_{NL} - V_{FL}}{V_{NL}} = \frac{(9 - 7.5)\ V}{9\ V}$$

that is, 0.17 per unit (or more typically, no units).
Expressed as a percentage, the regulation is

$$0.17 \times 100 = 17\%$$

(b) What would the regulation be if there was no change in the output voltage following the change in load current as in (a) above?
With no change in output voltage following the change in load current,

$$\text{regulation} = \frac{0\ V}{9\ V} = 0 = 0\%$$

Important point

- The lower the figure for regulation, the better the power supply.

The question remains – what causes the output voltage to fall?
Practical Exercise 2.4, which follows, may help to provide the answer.

Practical Exercise 2.4

To investigate the regulation of a power supply.
For this exercise you will need the following components and equipment:

1 – stabilized DC power supply (12 V, 1 A)
1 – resistor (10 Ω, 2 W)
1 – variable resistor (100 Ω, 2 W)
1 – DC ammeter
1 – DC voltmeter

Connect up the circuit shown in Figure 2.26.
The resistor 'r' is shown enclosed in the dotted box as 'belonging' to the power supply and represents the internal resistance of this supply.

Procedure

1 Before connecting R_L, set the output voltage to 12 V. This will be the no-load voltage (V_{NL}).
2 Set R_L to its maximum value and connect it into the circuit. Note the value of load current (I_L) and output voltage (V_L).

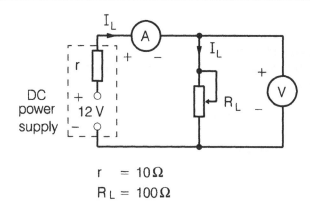

r $= 10\,\Omega$
$R_L = 100\,\Omega$

Figure 2.26 *Regulation: circuit diagram for Practical Exercise 2.4*

3 Adjust the value of R_L to give, say, five different values of I_L up to a maximum of 0.5 A, noting the corresponding values of V_L.
4 Complete the table shown in Figure 2.27.

Load voltage (V)						
Load current (A)	0					0.5

Figure 2.27 *Table of results for Practical Exercise 2.4*

5 Plot a graph of output voltage (vertically) against load current. A typical graph was shown in Figure 2.25.
6 At $I_L = 0.5$ A measure the fall in output voltage and hence calculate the regulation for this arrangement.
7 Remove resistor 'r' from the circuit and repeat procedures 2 to 5. The resulting graph should show little (or no) change in output voltage, thus giving a very low (or zero) value for regulation.

Conclusions

1 What is the cause of the output voltage decreasing with increasing load?
2 Obtain the manufacturer's specification document for the power supply used for this practical exercise and note the stated value for the following parameters:

(a) line regulation   (b) load regulation   (c) output impedance.

Questions

2.12 Two otherwise identical power supplies A and B have internal resistances of 1 Ω and 1.5 Ω respectively. State which of these supplies will have the better (lower) regulation.

2.13 The output voltage of a power supply falls from 30 V on no load to 25 V on full load. Calculate the percentage regulation.

2.14 A 12 V, 2 A power supply has an internal resistance of 3 Ω. Calculate the percentage regulation at full load.

Referring again to Figure 2.26, we can use DC circuit analysis to show that:

The voltage drop across 'internal resistor' r is given by

$V_r = I_L \times r$, while the load voltage V_L is given by

$$V_L = V_S - V_r$$
$$ = V_S - I_L r$$

Thus V_L will decrease as I_L increases, as the graph in Figure 2.25 shows.

The regulation of a power supply can be improved by the use of various voltage stabilization methods.

These methods include the use of another diode called the **zener diode** which will now be considered before regulation can be continued.

The zener diode

The zener diode is a silicon junction diode designed to break down in the reverse direction, at a **particular** voltage. Unlike the ordinary diode, where breakdown **must** be avoided, the zener diode is used specifically for this effect. The diagram in Figure. 2.28 shows a typical current–voltage characteristic.

In the forward direction the characteristic is like that of an ordinary diode.

In the reverse direction, little or no current will flow until a certain voltage, called the zener **breakdown voltage**, is reached.

The circuit symbol and cathode identification for the zener diode is shown in Figure 2.29.

Important points

- The zener diode is **always** used in the reverse biased condition.

- At breakdown, the voltage across the zener diode remains almost constant for a wide range of reverse currents. This breakdown voltage is related to a particular value of reverse current (e.g. 5 mA) and temperature (25° C).

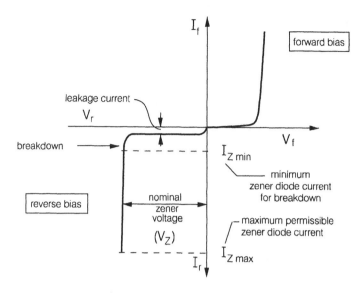

Figure 2.28 *The zener diode characteristic*

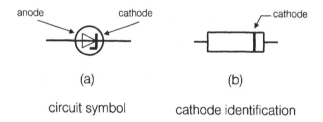

(a) | (b)
circuit symbol | cathode identification

Figure 2.29 *The zener diode: circuit symbol and cathode identification*

- The reverse current must be limited to a certain safe value by means of an external resistor.

- Zener diodes are specified in terms of their **breakdown** voltage and **power** rating, and are used as a voltage **regulator** or as a voltage **reference**.

- Zener diodes are available within the E12 and E24 series of preferred values of voltage and for various power ratings.

Example 2

A certain zener diode is specified as 10 V 500 mW. This means that its nominal breakdown voltage is 10 V with a power rating of 500 mW, which must **not** be exceeded.

The reverse current must therefore be limited to a maximum value, calculated from

$$I = \frac{P}{V} \text{ which gives}$$

$$I_r \max = \frac{500 \times 10^{-3} \text{ W}}{10 \text{ V}}$$

$$= 50 \times 10^{-3} \text{ A}$$

$$= 50 \text{ mA}$$

Example 3

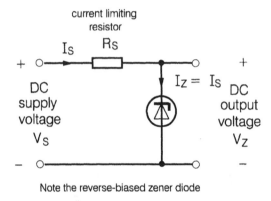

current limiting
resistor

Note the reverse-biased zener diode

Figure 2.30   *The zener diode: basic circuit diagram with current limiting resistor*

See Figure 2.30.

The value of the current limiting resistor to be connected in series with the zener diode is calculated as follows:

Voltage across series resistor $= (V_S - V_Z)$ where V_Z is the breakdown voltage.

If the maximum zener diode current $= I_Z \max$, then

$$R_S = \frac{(V_S - V_Z)}{I_Z \max}$$

Using the diode specified in Example 1, and a supply voltage of 25 V, we can calculate the value of the series resistor as

$$R_S = \frac{(V_S - V_Z)}{I_Z \max}$$

$$= \frac{(25 - 10) \text{ V}}{50 \times 10^{-3} \text{ A}}$$

$$= \frac{15 \text{ V}}{0.05 \text{ A}}$$

$$= 300 \text{ }\Omega$$

The preferred value in the E12 series would be 330 Ω, which would actually limit the zener diode current to a (safer) value of

$$\frac{15}{330} \text{ A or 45 mA}$$

The necessary power rating of this series resistor can be calculated from either $P = VI$ or $P = I^2R$.

Let us use $P = VI$, which gives a value of

$$15 \text{ (V)} \times 45 \times 10^{-3} \text{ (A) watts, or 675 mW}$$

A 1 W resistor would be needed.

Practical Exercise 2.5

To obtain the forward and reverse characteristics of a zener diode.
   For this exercise you will need the following components and equipment:

1 – zener diode (4.7 V, 500 mW)
1 – resistor (1 kΩ)
1 – DC power supply (variable from 0 to +25 V)
1 – DC ammeter
1 – DC voltmeter

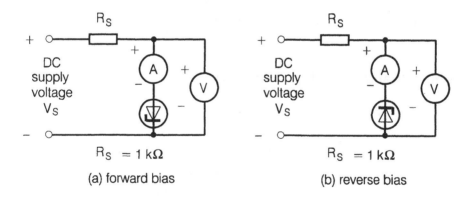

$R_S = 1 \text{ k}\Omega$

(a) forward bias

$R_S = 1 \text{ k}\Omega$

(b) reverse bias

Figure 2.31 *The zener diode characteristics: circuit diagrams for Practical Exercise 2.5*

Procedure

Figure 2.31(a) shows the forward biased zener diode.
1 Record the values of diode current (I_f) and voltage (V_f) as the supply voltage is increased from zero to +25 V.
   Figure 2.31(b) shows the reverse biased zener diode.

Continued on p. 34

Practical Exercise 2.5 *(Continued)*

2 Record the values of diode current (I_r) and voltage (V_z) as the supply voltage is increased from zero to +25 V.
3 Plot a graph of the above results, which should resemble the characteristic shown in Figure 2.28.

Conclusions

1 Examine the reverse characteristic and estimate the change in voltage across the diode as the current through it varies from say 1 mA to the maximum value.
2 At what value of current is the voltage across the diode equal to its nominal breakdown value?
3 Consult the supplier's catalogue and list the available range of power ratings. Notice also the preferred series of voltage values.

Voltage stabilization (regulation)

This is a method used to attempt to maintain a constant output voltage over the expected range of load currents and thus improve the regulation. There are various ways of achieving this, with success being largely related to the complexity of the circuit.

The shunt (parallel) voltage regulator

The basic circuit is shown in Figure 2.32.

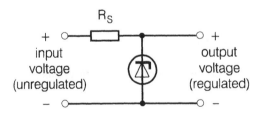

Figure 2.32   *The shunt voltage regulator: basic circuit diagram*

The purpose of any power supply is to provide a certain voltage at a certain current to an external load. Assuming a constant load voltage (which after all is what voltage regulation is about!) it will be necessary to know to what extent both the load current and the supply voltage is expected to vary. The action will be studied by means of the following practical exercise.

Practical Exercise 2.6

To investigate the use of a zener diode as a simple shunt voltage regulator. For this exercise you will need the following components and equipment:

1 – zener diode (4.7 V, 500 mW)
1 – resistor (330 Ω, 0.5 W)
1 – resistor (100 Ω, 0.5 W)
1 – variable resistor (1 kΩ)
1 – DC power supply (variable from +15 V to +25 V)
3 – DC ammeter
1 – DC voltmeter

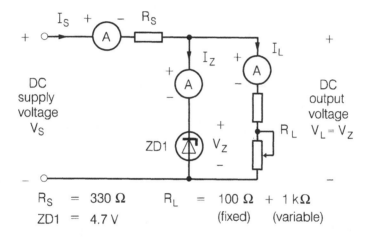

$$R_S = 330\ \Omega \qquad R_L = 100\ \Omega + 1\ k\Omega$$
$$ZD1 = 4.7\ V \qquad \text{(fixed)} \quad \text{(variable)}$$

Figure 2.33 *The shunt voltage regulator: circuit diagram for Practical Exercise 2.6*

Connect up the circuit shown in Figure 2.33.

Procedure (a) No load

1 Leave the load resistors (R_L) unconnected.
2 Vary the input supply voltage (V_S) from +15 V to +25 V and note the corresponding values of the voltage across the zener diode (V_Z) which will be taken to be the same as the output load voltage (V_L). (There will be a small voltage drop across the diode current ammeter which will be disregarded.)
3 Enter the results in table (a) in Figure 2.34.

Procedure (b) Variation in load current

1 Set the supply voltage to +20 V and connect the load resistors.

Continued on p. 36

Practical Exercise 2.6 (*Continued*)

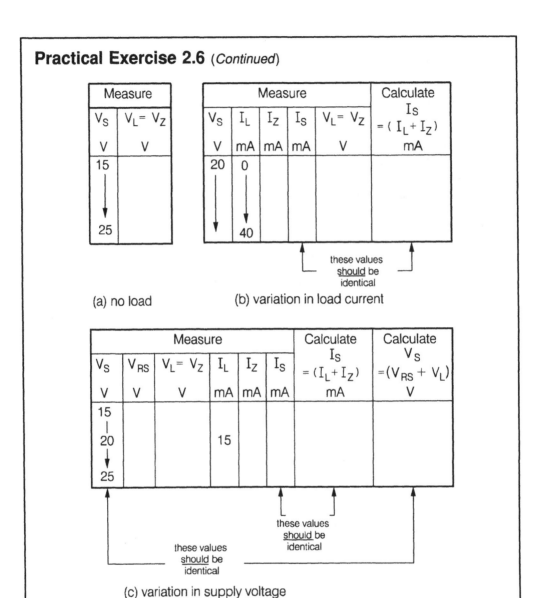

(a) no load   (b) variation in load current

(c) variation in supply voltage

Figure 2.34 *Table of results for Practical Exercise 2.6*

2 Increase the load current from zero to 40 mA maximum and note the corresponding values of:

(a) the zener diode current (I_Z)
(b) the supply current (I_S)
(c) the output load voltage (V_L).

3 Enter the results in table (b) in Figure 2.34.

Procedure (c) Variation in supply voltage

1 With a supply voltage of +20 V, adjust the load resistor to give a load current of 15 mA.
2 Vary the supply voltage from +15 V to +25 V and note the corresponding values of:
   (a) the voltage across the series resistor R_S(V_{RS})
   (b) the output load voltage
   (c) the load, zener diode and supply currents.
3 Enter the results in table (c) in Figure 2.34.
4 From the results in Figure 2.34(b), calculate:

$$\text{output resistance} = \frac{\text{change of output voltage}}{\text{change of output current}}$$

5 From the results in Figure 2.34(c), calculate:

$$\text{stabilization ratio} = \frac{\text{change of output voltage}}{\text{change of input (supply) voltage}}$$

Conclusions

For the following statements 1 and 2, choose the correct word within each bracket:
1 For a constant supply voltage, when the load current increases, the zener diode current *(increases/decreases)*, and the supply current *(increases/decreases)*. Over this variation of load current, the load voltage *(increases/decreases/remains virtually constant)*.
2 For a constant load current, when the supply voltage increases, the supply current *(increases/decreases)*, and the zener diode current *(increases/decreases)*. Over this variation of supply voltage, the load voltage *(increases/decreases/remains virtually constant)*.
3 What happens to the load voltage when the zener diode current falls to below a certain value? How effective is this circuit then as a regulator? What would you consider to be the minimum and maximum values of current for the particular diode used, in order to provide effective regulation?

The shunt regulator can be considered further by way of an example.

Example 4

(a) A zener diode and a resistor are connected in series to provide an 18 V stabilized output voltage from a 28 V supply. The circuit diagram is shown in Figure 2.35.
   If the load resistance is 450 Ω and the current through the diode is 10 mA, determine the value of the series resistor.
   $V_S = 28$ V, $V_L = V_Z = 18$ V, $R_L = 450$ Ω and $I_Z = 10$ mA.

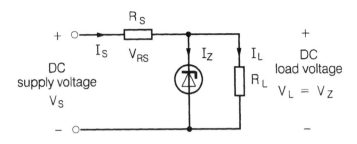

Figure 2.35   *Circuit diagram for Example 4*

Thus we have

$$V_{RS} = V_S - V_Z$$

$$= (28 - 18) \text{ V} = 10 \text{ V}$$

$$I_L = \frac{V_L}{R_L}$$

$$= \frac{18 \text{ V}}{450 \text{ }\Omega} = 40 \text{ mA}$$

$$I_S = I_Z + I_L$$

$$= (10 + 40) \text{ mA} = 50 \text{ mA}$$

$$R_S = \frac{V_{RS}}{I_S}$$

$$= \frac{10 \text{ V}}{50 \text{ mA}} = 200 \text{ }\Omega$$

(b) Calculate the diode current if the load resistance increases to 900 Ω.
    With $R_L = 900$ Ω, and assuming no change in V_L

then

$$I_L = \frac{18 \text{ V}}{900 \text{ }\Omega} = 20 \text{ mA}$$

Since

$$I_S = I_Z + I_L$$

then

$$I_Z = I_S - I_L$$

$$= (50 - 20) \text{ mA} = 30 \text{ mA}$$

(c) The zener diode current range for satisfactory operation is from 1 mA to 60 mA.
    Calculate the maximum and minimum value that the supply voltage can have for
    stabilization to be effective with a load resistance of 450 Ω. Assume that the zener
    diode voltage remains at 18 V for this range of zener diode current.

$$V_Z = 18 \text{ V and } I_L = 40 \text{ mA}$$

The minimum value of the supply, $V_S$min, is found by taking the minimum value of I_Z.

Thus, when $I_Z = 1$ mA

$$I_S = (1 + 40) \text{ mA} = 41 \text{ mA}$$

$$V_{RS} = I_S R_S$$

$$= 41 \text{ mA} \times 200 \text{ } \Omega = 8.2 \text{ V}$$

Giving

$$V_S\text{min} = V_{RS} + V_Z$$

$$= (8.2 + 18) \text{ V} = 26.2 \text{ V}$$

The maximum value of the supply, $V_S$max, is found by taking the maximum value of I_Z.

Thus, when $I_Z = 60$ mA

$$I_S = (60 + 40) \text{ mA} = 100 \text{ mA}$$

$$V_{RS} = I_S R_S$$

$$= 100 \text{ mA} \times 200 \text{ } \Omega = 20 \text{ V}$$

Giving

$$V_S\text{max} = V_{RS} + V_Z$$

$$= (20 + 18) \text{ V} = 38 \text{ V}$$

To summarize this part of the answer, we have calculated that the supply voltage can vary between 26.2 V and 38 V, and still maintain the output load voltage at 18 V $\pm$ 0 V.

In practice we must acknowledge that the voltage across the zener diode **will vary** with the current through it. This will be evident from either

(i) the reverse bias characteristic, which will then have a slope, that is, it will not be parallel to the I_r axis, or

(ii) a manufacturer's data sheet value for **slope resistance**.

Slope resistance of a zener diode (Ω) is given as

$$\frac{\text{change in voltage across the diode (V)}}{\text{change in current through the diode (A)}}$$

Note the use of basic units in the above expression as a matter of correctness. A glance at a catalogue will reveal values for slope resistance ranging from a few ohms to several hundred ohms.

To be more specific, an 18 V, 1.3 W zener diode is quoted as having a slope resistance of 20 Ω. Putting this value in the above expression gives

$$\text{slope resistance } = 20 \text{ } \Omega = \frac{20 \text{ V}}{1 \text{ A}} \text{ using basic units!}$$

In terms of the quantities actually involved, it comes down to

$$\text{slope resistance} \ = \ \frac{20 \text{ mV}}{1 \text{ mA}} \quad (= 20 \ \Omega)$$

Translating, this means that the voltage across the zener diode will change by 20 mV, for every 1 mA change in current through it.

Incorporating this detail of the diode in the above example, part (c) would typically read as follows:

> The zener diode current range for satisfactory operation is from 1 mA to 60 mA. Calculate the maximum and minimum value that the supply voltage can have for stabilization to be effective with a load resistance of 450 Ω. The slope resistance of the diode is 20 Ω. Calculate the corresponding maximum and minimum value of the load voltage.

The calculations are **exactly** as before, giving us **exactly** the same outcome so far, on one condition, which will be mentioned in a moment.

The information from part (a) gave

$$V_S = 28 \text{ V}, \ \text{with} \ I_Z = 10 \text{ mA at } V_Z = 18 \text{ V}$$

We now need to use the slope resistance detail to calculate any change in V_Z following any change in I_Z.

$V_S\text{min} = 26.2$ V gives $I_Z = 1$ mA, a change in I_Z of -9 mA.

The corresponding change in V_Z is -9 mA $\times$ 20 mV/mA, that is, -180 mV, or 0.180 V.

The load voltage $V_L = (18 - 0.180)$ V $= 17.82$ V

A similar calculation for $V_S\text{max} = 38$ V would show a change in I_Z of $+50$ mA, a change in V_Z of $+1$ V, giving a load voltage of 19 V. The revised summary is therefore that a variation in the supply voltage from 26.2 V to 38 V will give a variation in load voltage from 17.82 V to 19 V.

The condition referred to is that all calculations now assume that the load current remains at 40 mA, which of course it will not if the load voltage changes. For example, a load voltage of 19 V would mean a load current of 42.2 mA. However, to re-cycle the calculations at this stage in the hope of achieving greater accuracy would be difficult if not impossible, but certainly counter-productive.

One final point: the maximum current expected through the 200 Ω series resistor is 100 mA or 0.1 A. Its necessary minimum power rating is therefore given by $(0.1)^2 \times$ 200 watts or 2 W.

Questions

2.15 A zener diode, rated at 10 V, 1 W, is used with a supply whose voltage can vary from 13 V to 17 V. Calculate the minimum value of the series resistor and its necessary power rating.

2.16 The following data refers to a 400 mW zener diode:

nominal voltage at 5 mA = 10 V

slope resistance = 10 Ω

Sketch a circuit diagram showing how this diode may be used to supply a load current of 15 mA at an almost constant voltage of +10 V, from a DC supply of +20 V.
Calculate (a) the value of the series resistor
           (b) the change in load voltage for a change in supply voltage of ±2 V.

The series voltage regulator

The following two practical exercises will show different degrees of sophistication in providing series voltage regulation. Each circuit contains transistor circuitry which is yet to be covered. It is suggested that the exercises be worked through in turn, the results taken and the requested observations made and recorded, in readiness for a fuller explanation in due course.

Practical Exercise 2.7

To investigate the action of the simple series voltage regulator.
  For this exercise you will need the following components and equipment:

1 – zener diode (4.7 V, 500 mW)
1 – resistor (470 Ω, 0.5 W)
1 – resistor (100 Ω, 0.5 W)
1 – variable resistor (1 kΩ)
1 – DC power supply (variable from +15 V to +25 V)
1 – npn transistor (2N3053)
1 – DC voltmeter

  Connect up the circuit shown in Figure 2.36. The pin connection diagram for the 2N3053 transistor is shown in Figure 2.37.

Procedure (a) No load

1 Leave the load resistors (R_L) unconnected.
2 Vary the input supply voltage (V_S) from +15 V to +25 V and note the corresponding values of:
  (a) the voltage across the zener diode (V_Z)
  (b) the output load voltage (V_L).

Continued on p. 42

Practical Exercise 2.7 *(Continued)*

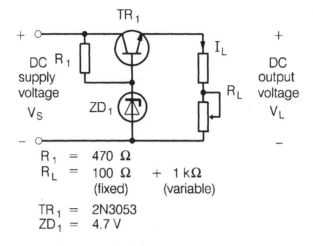

$$R_1 = 470 \ \Omega$$
$$R_L = 100 \ \Omega \quad + \quad 1 \ k\Omega$$
$$\text{(fixed)} \qquad \text{(variable)}$$

$$TR_1 = 2N3053$$
$$ZD_1 = 4.7 \ V$$

Figure 2.36 *The series voltage regulator: circuit diagram for Practical Exercise 2.7*

underside view

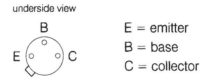

E = emitter
B = base
C = collector

Figure 2.37 *The 2N3053 transistor: pin connections*

3 Enter the results in table (a) in Figure 2.38.

Procedure (b) Variation in load current

1 Set the supply voltage to +20 V and connect the load resistors.
2 Adjust the variable resistor from maximum to minimum value to give, say, four different load currents. Note the corresponding values of:
   (a) voltage across the zener diode (V_Z)
   (b) output load voltage (V_L).
3 Enter the results in table (b) in Figure 2.38.

Procedure (c) Variation in supply voltage

1 With a supply voltage of +20 V, set the variable load resistor to about half-way.

2 Vary the supply voltage from +15 V to +25 V and note the corresponding values of:

(a) the voltage across the zener diode (V_Z)

(b) the output load voltage (V_L).

3 Enter the results in table (c) in Figure 2.38.

Measure			Calculate	
V_S	V_Z	V_L	(V_Z-V_L)	(V_S-V_L)
V	V	V	V	V
15				
↓				
25				

(a) no load

Measure			Calculate	
V_S	V_Z	V_L	(V_Z-V_L)	(V_S-V_L)
V	V	V	V	V
20				
↓				

(b) variation in load current

Measure			Calculate	
V_S	V_Z	V_L	(V_Z-V_L)	(V_S-V_L)
V	V	V	V	V
15				
↓				
25				

(c) variation in supply voltage

Figure 2.38 *Table of results for Practical Exercise 2.7*

4 From the results in Figure 2.38, calculate:

(a) output resistance

(b) stabilization ratio.

Conclusions

1 From the results in procedure (b)1 calculate the power being dissipated in the series regulator transistor TR_1 when the load current I_L is a maximum. Compare this value with the specification given in the manufacturer's data sheet.

Hint. Power in $TR_1 = 'V \times I'$, where $V = (V_S - V_L)$

and $I = I_L$

Continued on p. 44

Practical Exercise 2.7 (*Continued*)

2 Note the values of the voltage given from $(V_L - V_Z)$. Is this voltage difference close to the expected voltage drop across a forward biased diode?
3 From the results in Procedure (c)2 comment on the change in voltage across the zener diode as the input voltage is varied. Is this change reasonable?

Important point

- The load voltage will **follow** the zener diode voltage.

Practical Exercise 2.8

To investigate the action of the series voltage regulator using feedback and amplification.
   For this exercise you will need the following components and equipment:

1 – zener diode (4.7 V, 500 mW)
1 – npn transistor (2N3053)
1 – npn transistor (BC109)
1 – resistor (470 Ω, 1 W)
3 – resistor (1 kΩ, 0.5 W)
2 – resistor (10 kΩ)
2 – resistor (100 Ω, 2 W)
1 – variable resistor (500 Ω, 2 W)
1 – potentiometer (1 kΩ)
1 – DC power supply (variable from +15 V to +25 V)
1 – DC voltmeter

Connect up the circuit shown in Figure 2.39. The pin connection diagram for the 2N3053 transistor is given in Figure 2.37 and for the BC109 transistor in Figure 2.40.

Procedure (a) No load

1 Leave the load resistors (R_L) unconnected.
2 Vary the input supply voltage (V_S) from +15 V to +25 V and note the corresponding values of:
   (a) the output load voltage (V_L)
   (b) the voltage across resistor R_4 (V_{R4})
   (c) the voltage across the zener diode (V_Z).

3 Enter the results in table (a) in Figure 2.41.

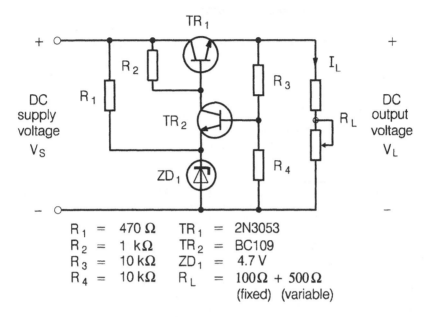

$$
\begin{array}{llll}
R_1 & = & 470\ \Omega & TR_1 & = & 2N3053 \\
R_2 & = & 1\ k\Omega & TR_2 & = & BC109 \\
R_3 & = & 10\ k\Omega & ZD_1 & = & 4.7\ V \\
R_4 & = & 10\ k\Omega & R_L & = & 100\,\Omega + 500\,\Omega
\end{array}
$$

(fixed)   (variable)

Figure 2.39 *The series voltage regulator using feedback and amplification, fixed output: circuit diagram for Practical Exercise 2.8 (Procedures (a), (b) and (c))*

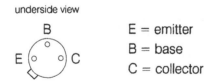

underside view

E = emitter
B = base
C = collector

Figure 2.40 *The BC109 transistor: pin connections*

Procedure (b) Variation in load current

1 Set the supply voltage to +20 V.
2 Adjust the variable load resistor from maximum to minimum value to give, say, four different load currents. Note the corresponding values of:
   (a) the output load voltage (V_L)
   (b) the voltage across resistor R_4 (V_{R4})
   (c) the voltage across the zener diode (V_Z).
3 Enter the results in table (b) in Figure 2.41.

Continued on p. 46

Practical Exercise 2.8 *(Continued)*

Procedure (c) Variation in supply voltage

1 With a supply voltage of +20 V, set the variable load resistor to about half-way.
2 Vary the supply voltage from +15 V to +25 V and note the corresponding values of the voltage:
 (a) at the output (V_L)
 (b) across the resistor R_4 (V_{R4})
 (c) across the zener diode (V_Z).
3 Enter the results in table (c) in Figure 2.41.

Measure				Calculate
V_S	V_L	V_{R4}	V_Z	$(V_{R4} - V_Z)$
V	V	V	V	V
15				
$\downarrow$				
25				

(a) no load

Measure				Calculate
V_S	V_L	V_{R4}	V_Z	$(V_{R4} - V_Z)$
V	V	V	V	V
20				
$\downarrow$				

(b) variation in load current

Measure				Calculate
V_S	V_L	V_{R4}	V_Z	$(V_{R4} - V_Z)$
V	V	V	V	V
15				
$\downarrow$				
25				

(c) variation in supply voltage

Figure 2.41 *Table of results for Practical Exercise 2.8 (Procedures (a), (b) and (c))*

4 From the results in Figure 2.41, calculate:
 (a) output resistance
 (b) stabilization ratio.

Conclusions

1 Are the values of the voltages given by ($V_{R4} - V_Z$) close to the expected voltage drop across a forward biased silicon diode?
2 Is this voltage regulator any improvement on the one in Practical Exercise 2.7?
   Hint. It **will** be if the values for stabilization ratio and output resistance are lower.

Procedure (d)

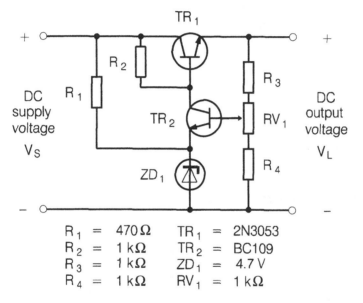

$$R_1 = 470\,\Omega \qquad TR_1 = 2N3053$$
$$R_2 = 1\,k\Omega \qquad TR_2 = BC109$$
$$R_3 = 1\,k\Omega \qquad ZD_1 = 4.7\,V$$
$$R_4 = 1\,k\Omega \qquad RV_1 = 1\,k\Omega$$

Figure 2.42 *The series voltage regulator using feedback and amplification, variable output: circuit diagram for Practical Exercise 2.8 (Procedure (d))*

1 Modify the circuit to that shown in Figure 2.42.
2 Investigate the effect on the output voltage of varying the position of the slider of the potentiometer RV_1. Record the minimum and maximum values of this output voltage for various values of supply voltage.

Important point

- The series transistor in this type of regulator is acting as a variable resistor. The voltage drop across this resistor will vary, in order to compensate for any tendency for change in the output load voltage.

Integrated circuit voltage regulators

There are a range of integrated circuits as voltage regulators which, to a large extent, have made discrete components obsolete. They are available for a range of fixed positive and negative output voltages for various output currents and also for variable voltage outputs (mostly positive).

The final practical exercise for this chapter will investigate the use of both fixed and variable voltage regulators.

Practical Exercise 2.9

To investigate the action of an integrated circuit voltage regulator.
For this exercise you will need the following components and equipment:

1 – fixed voltage regulator ic (7812)
1 – variable voltage regulator ic (317L)
1 – DC power supply (variable from 0 V to +25 V)
2 – capacitor (100 nF)
1 – capacitor (10 µF, 35 V)
1 – resistor (100 Ω, 0.5 W)
1 – potentiometer (1 kΩ, 1 W)
1 – DC voltmeter

Procedure (a) Fixed output voltage

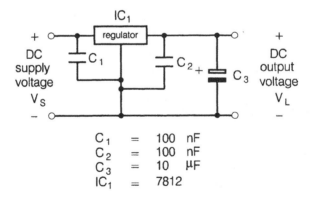

$$C_1 = 100 \ nF$$
$$C_2 = 100 \ nF$$
$$C_3 = 10 \ µF$$
$$IC_1 = 7812$$

Figure 2.43 *The integrated circuit voltage regulator, fixed output: Circuit diagram for Practical Exercise 2.9 (Procedure (a))*

The circuit is shown in Figure 2.43 and the pin connection diagram for the 7812 ic in Figure 2.44.

1 Vary the DC supply voltage in steps from 10 V to 25 V and note the corresponding DC output voltages.

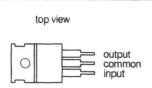

top view

output
common
input

Figure 2.44 *The 78 series integrated circuit voltage regulator: pin connections*

Conclusions

From your results, state:

1 The nominal output voltage of this regulator (5 V, 9 V, 12 V, 15 V or 24 V).
2 The minimum supply voltage that will provide this output voltage.

Procedure (b) Variable output voltage

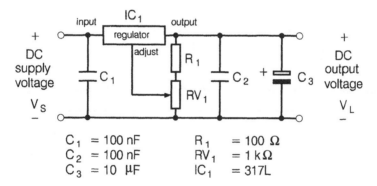

C_1 = 100 nF $\qquad$ R_1 = 100 Ω
C_2 = 100 nF $\qquad$ RV_1 = 1 kΩ
C_3 = 10 μF $\qquad$ IC_1 = 317L

Figure 2.45 *The integrated circuit voltage regulator, variable output: circuit diagram for Practical Exercise 2.9 (Procedure (b))*

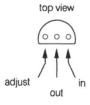

top view

adjust $\qquad$ in
out

Figure 2.46 *The 317L integrated circuit voltage regulator: pin connections*

The circuit is shown in Figure 2.45 and the pin connection diagram for the 317L ic in Figure 2.46.

Continued on p. 50

Practical Exercise 2.9 *(Continued)*

1 Set the DC supply voltage, in turn, to 2 V, 10 V, 15 V and 25 V and for each value vary RV_1 to find the resulting variation in the DC output voltage.

Conclusions

From your results, state:

1 The range in output voltage achieved with a 2 V to 25 V variation in the supply voltage.
2 Consult the supplier's catalogues and complete the table in Figure 2.47.

TYPE	78L05	78M12	7815	7912	317L	337T	L200
Fixed/variable							
Output voltage							
Input voltage range							
Output current							
Line regulation							
Load regulation							
Output resistance							

Figure 2.47   *Integrated circuit voltage regulators: table for Practical Exercise 2.9*

3 Small signal amplifiers

Amplifier gain

In general, the fundamental requirement of an amplifier is that the output signal should be an enlarged (amplified) version of the input signal. The definition of amplification, otherwise known as 'gain', is given by

$$\text{gain} = \frac{\text{output signal}}{\text{input signal}}$$

It should be obvious that a power supply is required in order to provide the necessary energy for this amplification. See Figure 3.1.

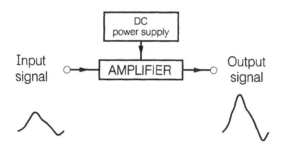

Figure 3.1 *Amplification*

The input signal can take many forms. In audio systems for example, it can be the voltage from a microphone, cassette deck or compact disc player, all of which will be of the order of millivolts.

Clearly, in order to provide sufficient power or 'drive' for a loudspeaker (say 50 W), amplification is needed. The block diagram of a typical system is shown in Figure 3.2.

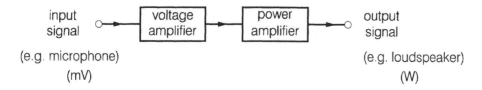

Figure 3.2 *Amplification block diagram*

The initial stages of such a system are called preamplifiers, containing the switching circuitry for each type of input together with the tone (equalization) controls. These are known as voltage amplifiers whose job is to raise the signal voltage to that level required by the power amplifiers.

Amplifiers are active devices and as such they provide gain, whether it be voltage, current or power. The distinction between voltage and power amplification comes about because the prime function of voltage amplifiers is to provide substantial voltage gain with very little power gain (there will be some). Power amplifiers on the other hand provide substantial power gain (= voltage gain times current gain).

Important points

- Voltage amplifiers are known as small-signal amplifiers because of the small changes in voltage and current which occur.

- Power amplifiers are large-signal for precisely the opposite reason.

- Some 'amplifiers' may be designed to provide unity gain but at the same time perform an alternative function.

The bipolar transistor

The diagram usually given to illustrate transistor action is that shown in Figure 3.3(a), which depicts the transistor as a pnp or npn sandwich. It must be stressed that the actual construction of the modern transistor bears zero resemblance to this diagram, which is therefore for convenience only.

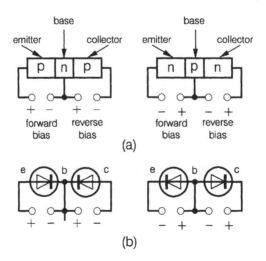

Figure 3.3 *The bipolar transistor: basic arrangement*

The transistor may be regarded as two diodes in a back-to-back situation, see Figure 3.3(b). The three electrodes or terminals are called **emitter** (e), **base** (b) and **collector** (c). The two diodes referred to are base–emitter and collector–base respectively.

Important points

- The base–emitter junction is **forward** biased, thus providing low resistance to current flow.

- The collector–base junction is **reversed** biased, providing high resistance to current flow.

- The transistor is bipolar because current in the device is due to both holes **and** electrons. This is a physical phenomenon for which lack of further explanation will not hinder the understanding of small signal amplification.

- The arrow head on the emitter shows the direction of conventional current flow.

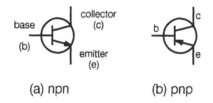

(a) npn   (b) pnp

Figure 3.4 *The bipolar transistor: circuit symbols*

The circuit symbols for the respective types are shown in Figure 3.4. The npn transistor, constructed from silicon, is more widely used than the pnp type (mainly germanium) and will be used throughout for all practical exercises.

Practical Exercise 3.1

To investigate the switching action of a bipolar transistor.
For this exercise you will need the following components and equipment:

1 – npn transistor (BC109)
1 – resistor (1 kΩ)
1 – variable resistor (1 MΩ)
1 – filament lamp (6 V)
1 – DC power supply (+5 V)
1 – DC ammeter
1 – DC voltmeter

Continued on p. 54

Practical Exercise 3.1 *(Continued)*

Procedure

1 Connect up the circuit shown in Figure 3.5.

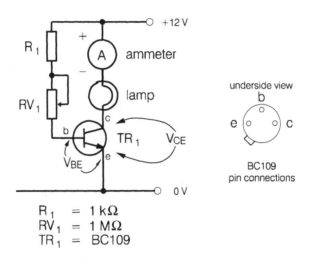

$$R_1 = 1\,k\Omega$$
$$RV_1 = 1\,M\Omega$$
$$TR_1 = BC109$$

Figure 3.5   *The transistor switch: circuit diagram for Practical Exercise 3.1*

2 Set RV_1 to its maximum value. The lamp should **not** be lit and the ammeter reading should be zero.

3 Decrease RV_1 so that the lamp begins to light. The transistor is now starting to conduct, that is, it is switching on. This will be shown by a reading on the ammeter.

   At the same time, you should also notice that:

   (a) the voltage V_{BE} increases from 0 V

   (b) the voltage V_{CE} decreases from +5 V.

4 Decrease RV_1 still further until the lamp brightness is a maximum and then measure the voltage:

   (a) V_{BE} (this should be approximately 0.6 V)

   (b) V_{CE} (this should be approximately zero)

   (c) across lamp (this should be approximately 12 V).

Important point

- At the maximum level of brightness the transistor is fully switched on, that is, fully conducting, or **saturated**.

5 Adjust RV_1 until V_{CE} equals 6 V. Measure:

(a) the base emitter voltage (V_{BE})
(b) the voltage across the lamp (V_L)
(c) the collector current, I_C
(d) the voltage across R_1.

6 Calculate:

(a) $I_B = \dfrac{V_{R1}}{R_1}$

(b) $h_{FE} = \dfrac{I_C}{I_B} = \dfrac{\text{output current}}{\text{input current}}$

7 Complete the table given in Figure 3.6.

		Measure				Calculate	
V_{CE}	V_{BE}	voltage across lamp V_L	I_C	V_{R1}	$I_B = \dfrac{V_{R1}}{R_1}$	$\dfrac{I_C}{I_B} = h_{FE}$	

Figure 3.6 *Table of results for Practical Exercise 3.1*

Important points

- The bipolar transistor is a current-operated device.

- The forward biased emitter-base junction provides a small current in the base which gives rise to a larger current in the collector. This is current amplification or **gain**.

- DC current gain, I_C/I_B is given the symbol 'h_{FE}'.

Methods of connection

There are three methods, each of which is defined in terms of the terminal or electrode which is common to **both** the input and output circuit. The characteristics of the transistor are quite different when used in each of these ways.

(a) Common emitter

The **basic** biasing arrangement is shown in Figure 3.7(a), which shows two separate DC supplies, one for each junction. The **practical** circuit in Figure 3.5 is also common emitter, and uses only one DC supply, with the forward bias for the base–emitter junction being obtained from the positive supply line through appropriate resistors.

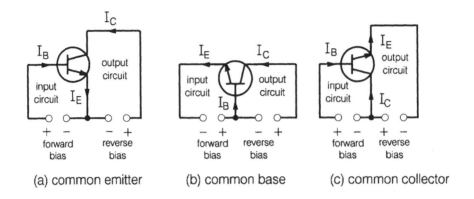

(a) common emitter   (b) common base   (c) common collector

Figure 3.7   *The bipolar transistor: methods of connection*

The value of I_C/I_B, or h_{FE}, is therefore the common emitter current gain.

Assuming a (typical) value of $h_{FE} = 100$ and $I_B = 10\ \mu A$, then from $h_{FE} = I_C/I_B$, we can see that

$$I_C = h_{FE} \times I_B = 100 \times 10\ \mu A = 1000\ \mu A, \text{ or } 1\ mA$$

From which, see Figure 3.7(a), it follows that

$$I_E = I_C + I_B = 1010\ \mu A$$

(b) Common base

Figure 3.7(b) shows the basic arrangement. It must be emphasized that this is the **same transistor**, connected differently.

The DC current gain in common base, known as h_{FB}, is given by

$$h_{FB} = \frac{\text{output current}}{\text{input current}} = \frac{I_C}{I_E}, \text{ thus giving } h_{FB} = \frac{1000\ \mu A}{1010\ \mu A}$$

$$= 0.99$$

(c) Common collector

Figure 3.7(c) shows the basic arrangement. Again, it is the **same** transistor, this time providing a DC current gain h_{FC}, given by

$$h_{FC} = \frac{\text{output current}}{\text{input current}} = \frac{I_E}{I_B}$$

$$= \frac{1010\ \mu A}{10\ \mu A} = 101$$

Important points

- A figure of less than unity would rule out the use of **common base** as a current amplifier. However, it has the characteristic of low input resistance and high output resistance which, together with power gain, makes it useful as an **impedance matching** device.

- The **common collector**, although having the highest value of current gain, is also used for **impedance matching**, possessing as it does, high input and low output resistance and power gain. Its alternative name is **emitter follower**.

- The most widely used arrangement is **common emitter**, which has current, voltage and power gain.

A comparison of typical transistor parameters is given in Figure 3.8.

	Common base	Common emitter	Common collector
Current gain	just less than unity	✓	✓
Voltage gain	✓	✓	just less than unity
Power gain	✓	highest	✓
Input resistance	low (Ω)	medium (kΩ)	high (MΩ)
Output resistance	high (MΩ)	medium (kΩ)	low (Ω)
Phase inversion	0	180	0
Application	rf amplifier	general	buffer

Figure 3.8 *Comparison of the three types of transistor amplifier*

Practical Exercise 3.2

To obtain the input, output and transfer characteristics of a common emitter transistor.

Continued on p. 58

Practical Exercise 3.2 *(Continued)*

For this exercise you will need the following components and equipment:

1 – npn transistor (BC109)
1 – DC supply (variable from 0 V to +12 V)
1 – DC supply (variable from 0 V to +1 V)
1 – DC microammeter
1 – DC ammeter
1 – DC voltmeter (preferably digital)

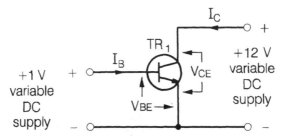

Figure 3.9 *Bipolar transistor characteristics: circuit diagram for Practical Exercise 3.2*

Make up the circuit shown in Figure 3.9, making sure for each part of the exercise that false readings are not obtained due to incorrect positioning of the meters. As a reminder, refer to Chapter 2, Figures 2.3(a) and 2.4(a).
The pin connection diagram for the BC109 transistor is given in Figure 3.5.

Procedure (a) Input characteristics

1 Set V_{CE} to +6 V and keep this constant.
2 Vary the +1 V supply from zero to give readings of I_B and V_{BE}, up to $I_B = 100 \ \mu A$.
3 Complete the table shown in Figure 3.11(a).
4 Plot the graph of I_B (vertically) against V_{BE}.
A typical characteristic is shown in Figure 3.10(a).

Procedure (b) Output characteristics

1 With $I_B = 0$, vary V_{CE} over the range 0 V to +12 V in steps and at each step note the value of I_C.
2 Repeat procedure 1 for values of I_B of:
   (a) 20 μA (b) 40 μA (c) 60 μA and (d) 80 μA
3 Complete the table shown in Figure 3.11(b).
4 Draw the graph of I_C (vertically) against V_{CE} for each value of I_B to give a family of characteristics.

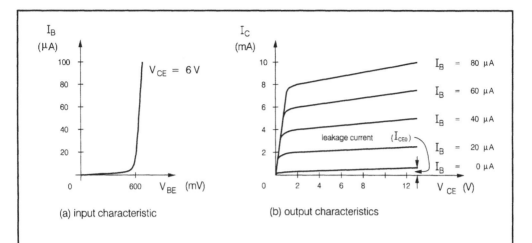

(a) input characteristic

(b) output characteristics

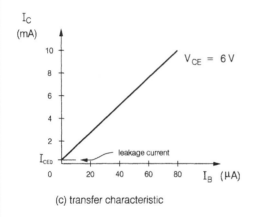

(c) transfer characteristic

Figure 3.10 *The common emitter bipolar transistor: typical input, output and transfer characteristics*

A typical set of output characteristics is shown in Figure 3.10(b).

5 From the graph, when $V_{CE} = 6$ V, estimate the value of I_C for $I_B = 40$ μA.

To do this, draw a vertical line starting from $V_{CE} = 6$ V, and at the intersection of this line with the curve of $I_B = 40$ μA, project horizontally to meet the I_C axis. Hence determine the value of the DC current gain at $V_{CE} = 6$ V.

Procedure (c) Transfer characteristics

1 Set V_{CE} to $+6$ V.
2 Vary I_B from 0 to 100 μA in steps and measure I_C.
3 Complete the table shown in Figure 3.11(c).
4 Draw the graph of I_C (vertically) against I_B.

A typical transfer characteristic is shown in Figure 3.10(c).

Continued on p. 60

Practical Exercise 3.2 (*Continued*)

5 From the graph, estimate the value of I_C when $I_B = 40\ \mu\text{A}$ and hence determine the value of DC current gain. This value should be the same as that found in procedure (b)5!

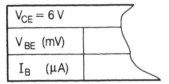

$V_{CE} = 6$ V	
V_{BE} (mV)	
I_B (μA)	

(a) input characteristic

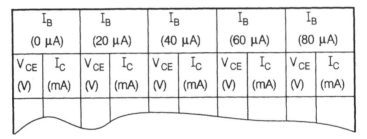

I_B (0 μA)		I_B (20 μA)		I_B (40 μA)		I_B (60 μA)		I_B (80 μA)	
V_{CE} (V)	I_C (mA)	V_{CE} (V)	I_C (mA)	V_{CE} (V)	I_C (mA)	V_{CE} (V)	I_C (mA)	V_{CE} (V)	I_C (mA)

(b) output characteristics

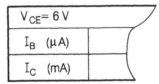

$V_{CE} = 6$ V	
I_B (μA)	
I_C (mA)	

(c) transfer characteristic

Figure 3.11   *Table of results for Practical Exercise 3.2*

Conclusion

1 Consult catalogue data to find the range of h_{FE} for the BC109 transistor, in each of the following categories: Group A, Group B, Group C.

Questions

3.1 A transistor has a base current of 10 μA and an h_{FE} of 250. Calculate the collector current.

3.2 A transistor with an h_{FE} of 400 has a collector current of 2 mA. Calculate the base current.

3.3 A transistor has a base current of 40 µA and a collector current of 5 mA. Calculate the DC current gain.

The bipolar transistor amplifier

The word 'transistor' originates from two words, **trans**fer and re**sistor**. Practical Exercise 3.1 will have shown that a small current flowing in a low resistance (base–emitter) junction can result in a much larger current (h_{FE} times as large) flowing in a high resistance (collector–base) junction. Thus the idea of transferring resistance is formed. Of much more importance is the fact that there will be substantial voltage and power gains due to this effect.

Let us have a look at the common emitter amplifier, one circuit for which is given in Figure 3.12.

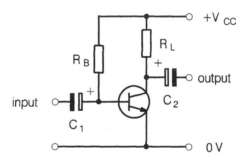

Figure 3.12 *The common emitter amplifier: basic circuit diagram*

Important points

- R_L is the **collector load** and provides the output voltage.

- R_B is the **biasing** resistor and sets the correct operating point for the transistor.

- C_1 and C_2 are AC signal coupling capacitors which act as a DC block at the input and output respectively.

- Biasing is the process whereby the base–emitter junction is forward biased and the collector–base junction reverse biased.

Practical Exercise 3.3

To investigate the action of a common emitter transistor amplifier.

Continued on p. 62

Practical Exercise 3.3 (*Continued*)

For this exercise you will need the following components and equipment:

1 – npn transistor (BC109, BC108, BC107)
1 – resistor (1 kΩ, 2.2 kΩ, 1 MΩ)
2 – capacitor (100 µF, 16 V)
1 – DC power supply (+12 V)
1 – audio frequency signal generator
1 – double beam cathode ray oscilloscope
1 – DC voltmeter

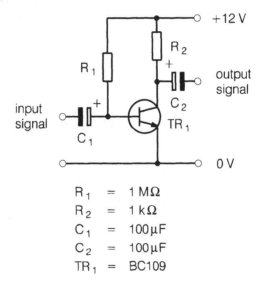

$$R_1 = 1\,\text{M}\Omega$$
$$R_2 = 1\,\text{k}\Omega$$
$$C_1 = 100\,\mu\text{F}$$
$$C_2 = 100\,\mu\text{F}$$
$$TR_1 = BC109$$

Figure 3.13 *The common emitter amplifier: circuit diagram for Practical Exercise 3.3*

Procedure

1 Connect up the circuit shown in Figure 3.13, leaving the signal generator and cathode ray oscilloscope unconnected for the moment. The pin connection diagram for the BC109 transistor is given in Figure 3.5.

Important point

- To check that the required **DC operating conditions** have been achieved, it is usual to measure the voltages at the base and collector with respect to the 0 V line (V_{BE} and V_{CE} respectively). If V_{BE} is of the order of 0.6 V, then the transistor is likely to be correctly biased, in which case, for best results, V_{CE} should be about 6 V. Other values for V_{CE} will be acceptable but there will then be limits to the maximum amplification available.

2 Measure the voltages V_{BE} and V_{CE}.

3 Connect the signal generator and set the frequency to 1 kHz with an output sinusoidal voltage of 10 mV peak-to-peak. Measure the output voltage, which should be sinusoidal.

4 Calculate the voltage gain (A_v) from

$$A_v = \frac{\text{output signal voltage}}{\text{input signal voltage}}$$

(use peak-to-peak values **for both**).

5 Display both input and output signal waveforms and note the phase difference between them.

6 Repeat procedures 3 and 4 with $R_1 = 2.2$ kΩ.

7 Complete the table shown in Figure 3.14.

Load resistor R_2	V_{BE}	V_{CE}	Input signal voltage	Output signal voltage	Voltage gain
1 kΩ					
2.2 kΩ					
Phase difference between output and input signals					

Figure 3.14 *Table of results for Practical Exercise 3.3*

Important points

- For **small-signal amplifiers** such as this, the current gain under signal conditions, known as AC current gain (h_{fe}, or A_i), can be assumed to be equal to the DC current gain. Thus, $h_{fe} = h_{FE}$.

- Power gain (A_p) = current gain (A_i) × voltage gain (A_v).

8 Use the DC measurements taken in procedure 2 ($R_L = 1$ kΩ) to calculate the value of h_{FE}, and hence h_{fe}, as follows:

$$V_{RB} = V_{CC} - V_{BE} \quad \text{and} \quad V_{RL} = V_{CC} - V_{CE}$$

$$I_B = \frac{V_{RB}}{R_B} \quad \text{and} \quad I_C = \frac{V_{RL}}{R_L},$$

Continued on p. 64

Practical Exercise 3.3 (*Continued*)

$$h_{FE} = h_{fe} = \frac{I_C}{I_B},$$

where $V_{CC} = 12$ V, $R_L = 1$ kΩ and $R_B = 1$ MΩ.

9 Calculate the power gain of this amplifier.
10 Repeat the voltage gain measurement using the BC107 and BC108 transistors. In addition, try a number of each of the three types to see if there are substantial differences in voltage gain.

Conclusions

1 What is the observed phase shift between output and input signal voltages?
2 Compare the measured voltage gains for each type of transistor.

Practical Exercise 3.4

To investigate the effect of varying the biasing conditions on the operation of a common emitter transistor amplifier.

For this exercise you will need the following components and equipment:

1 – npn transistor (BC109)
1 – resistor (1 kΩ, 4.7 kΩ, 1 MΩ)
1 – variable resistor (1 MΩ)
2 – capacitor (100 μF, 16 V)
1 – DC power supply (+12 V)
1 – audio frequency signal generator
1 – double beam cathode ray oscilloscope
1 – DC voltmeter

Procedure

1 Connect up the circuit of Figure 3.15. The pin connection diagram for the BC109 transistor is shown in Figure 3.5.
2 The signal generator output voltage should be set to 10 mV peak-to-peak at 1 kHz. Monitor the collector–emitter voltage (V_{CE}) with the voltmeter.
3 Adjust RV_1 to give $V_{CE} = 6$ V. Increase the input signal voltage to a point where the first signs of output signal distortion occur. This will most likely be seen as a flattening of the bottom part of the waveform.
4 Reduce the input signal voltage so that this waveform is now **just** not distorted and leave it set at this level.

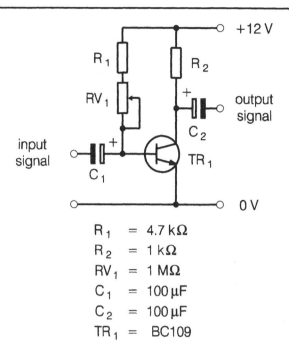

R_1 = 4.7 kΩ
R_2 = 1 kΩ
RV_1 = 1 MΩ
C_1 = 100 μF
C_2 = 100 μF
TR_1 = BC109

Figure 3.15 *The common emitter amplifier: circuit diagram for Practical Exercise 3.4*

Important point

- With the collector–emitter voltage equal to **one-half** of the supply, and the output signal **just** not distorted, the amplifier is giving optimum linear performance.

---

5 Adjust RV_1 so that V_{CE} falls to 4 V and note the shape of the output waveform. This is the result of **too much** bias current, which should cause the output waveform to be flattened on the bottom half.

6 Adjust RV_1 so that V_{CE} rises and again note the shape of the output waveform. This is the result of **too little** bias current, which should cause the output waveform to be rounded (rather than flattened) on the top half. This particular effect may be more noticeable if R_1 is changed to 1 MΩ.

Conclusions

1 What do you think is meant by the term 'linear amplifier'?
2 What would be the result of a Hi-Fi amplifier not maintaining linear operation?

The load line

The performance of the transistor amplifier can be examined by the use of a so-called **load line** drawn on the output characteristics.

Figure 3.16 shows the result of using the following steps to obtain the position of this load line.

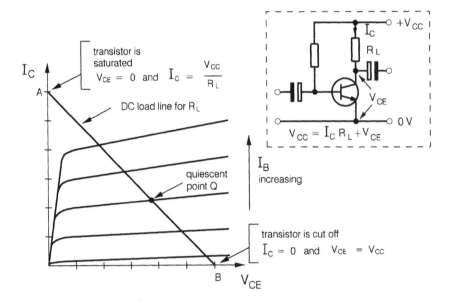

Figure 3.16 *The common emitter amplifier load line: (a) derivation*

(a) The DC condition is given by $V_{CC} = I_C R_L + V_{CE}$.
(b) When $V_{CE} = 0$ V, the transistor is **saturated**, and $I_C = V_{CC}/R_L$. This gives point A on the graph.
(c) When $I_C = 0$ mA, the transistor is **cut off**, and $V_{CE} = V_{CC}$. This gives point B on the graph.
(d) Joining the points A and B gives the **DC load line** for the particular value of collector load resistor R_L and supply voltage V_{CC}.

Important points

- The intersection of the **DC load line** with the required value of base bias current gives the operating point (P), or **quiescent point** (Q). Quiescent means quiet or 'no-signal' condition.

- See Figure 3.17.
  The values of I_C and V_{CE} where lines from the Q point intersect the respective axes are the quiescent or no-signal values.

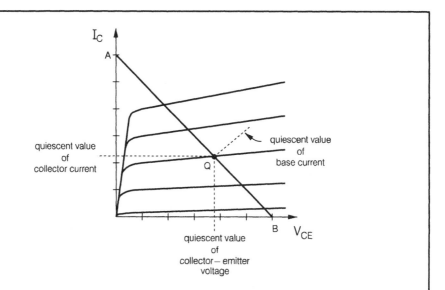

Figure 3.17 *The common emitter amplifier load line: (b) quiescent conditions*

- For this particular arrangement, the load line not only represents the DC conditions relating to the amplifier but also those for the AC signal. For this reason it is known as a DC and AC load line.

- See Figure 3.18.

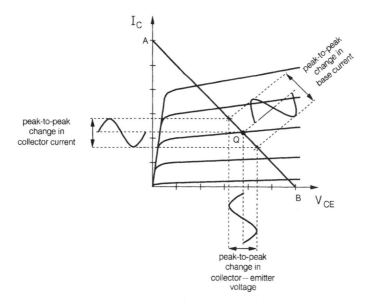

Figure 3.18 *The common emitter amplifier load signal: (c) signal conditions*

Continued on p. 68

Important points (*Continued*)

The application of an input signal voltage causes an input signal current to flow in the base. The base current, the collector current and the collector voltage will all vary above and below the quiescent value.

Example 1

A common emitter amplifier operates from a supply of +12 V with a load resistance of 1 kΩ and has the output characteristics shown in Figure 3.19. The required base bias current is 40 µA and the input resistance can be assumed to be 1.5 kΩ.

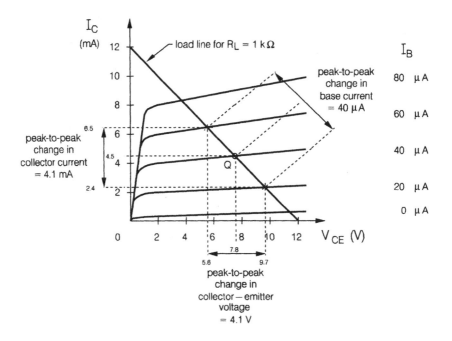

Figure 3.19   *Diagram for Example 1*

(a) The load resistance is 1 kΩ.
   Point A on the load line is given by

$$V_{CE} = 0 \text{ V and } I_C = \frac{V_{CC}}{R_L} = \frac{12 \text{ V}}{1 \text{ k}\Omega} = 12 \text{ mA}$$

   Point B is given by $V_{CE} = V_{CC} = 12$ V and $I_C = 0$ mA.
(b) For a base bias current of 40 µA, the quiescent point Q gives $I_C = 4.5$ mA.

   Hence the DC current gain $(h_{FE}) = \dfrac{4.5 \text{ mA}}{40 \text{ µA}} = 112.5$

(c) For an input signal current of 40 µA peak-to-peak, the change in base current is $(60 - 20)$ µA = 40 µA peak-to-peak.

(d) The resulting change in collector current is $(6.5 - 2.4)$ mA = 4.1 mA peak-to-peak, giving an

$$\text{AC current gain } (h_{fe}) \text{ of } \frac{4.1 \text{ mA}}{40 \text{ µA}} = 102.5$$

(almost the same as the DC current gain).

(e) The change in collector–emitter voltage is $(5.6 - 9.7)$ V $= (-)$ 4.1 V peak-to-peak.

(f) The input resistance is given as 1.5 kΩ. The input voltage must therefore be 40 µA × 1.5 kΩ, that is 60 mV, or 0.06 V peak-to-peak.

(g) The voltage gain for this amplifier is $(-)$ 4.1 V/0.06 V, that is, $(-)$ 68 times.

(h) The $(-)$ represents the 180 degrees phase difference between the output and input voltages. Notice from Figure 3.18 that when the input signal base current (and hence voltage) increases (goes positive), the output signal voltage between collector and emitter decreases (goes less positive).

Questions

3.4 The output characteristics of a common emitter transistor are linear between the points in the table:

V_{CE} (V)	2	25	
I_C (mA)	4.5	5.5	when $I_B =$ 50 µA
I_C (mA)	14.0	16.0	when $I_B =$ 150 µA
I_C (mA)	23.5	26.5	when $I_B =$ 250 µA

The transistor is used with a collector load resistance of 1 kΩ and a supply voltage of 30 V. The bias current is 150 µA.

(a) Plot the output characteristics.

(b) Construct the load line.

(c) Estimate the quiescent collector current.

(d) Calculate the DC current gain.

An input signal voltage of 800 mV peak-to-peak causes a base current swing of 200 µA peak-to-peak.

(e) Estimate the peak-to-peak collector current swing and hence calculate the AC current gain.

(f) Estimate the peak-to-peak collector-emitter voltage swing and hence calculate the voltage gain.

(g) Calculate the power gain from (e) and (f).

3.5 The output characteristics of a common emitter transistor are linear between the points in the table:

Continued on p. 70

Questions (*Continued*)

V_{CE} (V)	2	12	
I_C (mA)	2.4	3.4	when $I_B = 50$ μA
I_C (mA)	5.2	6.2	when $I_B = 100$ μA
I_C (mA)	7.8	8.8	when $I_B = 150$ μA
I_C (mA)	10.2	11.4	when $I_B = 200$ μA
I_C (mA)	12.2	14.0	when $I_B = 250$ μA

The transistor is used with a collector load resistance of 680 Ω and a supply voltage of 12 V. The bias current is 150 μA.

(a) Plot the output characteristics.

(b) Construct the load line and estimate the no-signal values of collector current and collector–emitter voltage.

(c) Estimate the voltage gain for an input current of 200 μA peak-to-peak through an input resistance of 1 kΩ.

(d) Notice that the spacing between the characteristics decreases at the higher levels of base current. What effect do you think this will have on the output voltage waveform?

(e) Estimate the base bias current necessary for the quiescent collector–emitter voltage to be 6 V (that is, one-half of the supply voltage).

Addition of an external load

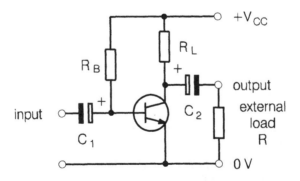

Figure 3.20 *The common emitter amplifier: addition of external load*

Figure 3.20 shows the previous amplifier of Figure 3.12 to which an external load (R) has been added.

This external load plays no part in determining the DC conditions of the amplifier because it is separated from the amplifier by a coupling capacitor behaving as a DC block. For the AC signal, to which the capacitor will normally be a low reactance path, this load resistor will effectively be in parallel with the collector load R_L.

As a result it will be necessary to draw separate DC and AC load lines in order to predict the performance of the amplifier.

The DC load line for R_L is drawn exactly as described earlier, and the no-signal values found. This line will have a gradient, or slope, equal to $-1/R_L$. To confirm this, look again at Figure 3.19. Since the line is 'going downhill', its gradient is negative, and the value of this gradient is equal to 4.1 mA/4.1 V or 1 mA/V. The load resistor R_L has a value of 1 kΩ, or 1 V/mA and the reciprocal of R_L is 1 mA/V.

The AC load line, from which all signal estimates must now be made, will pass through the previously determined quiescent point, but its gradient will be equal to $-1/R_L'$, where R_L' is the parallel combination of R_L and R.

If we let $R = 1$ kΩ, then $R_L' = 0.5$ kΩ, or 0.5 V/mA, giving a gradient for the AC load line of 2 mA/V. The result of drawing both load lines is shown in Figure 3.21.

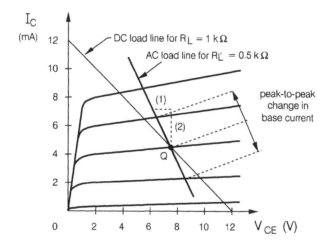

Figure 3.21 *The common emitter amplifier: DC and AC load lines*

Question

3.6 Refer to Question 3.4. This amplifier now feeds an external load resistance of 1 kΩ.

Add the AC load to your diagram and calculate the new value of voltage gain.

The importance of correct biasing

Figure 3.22 shows three different bias conditions.

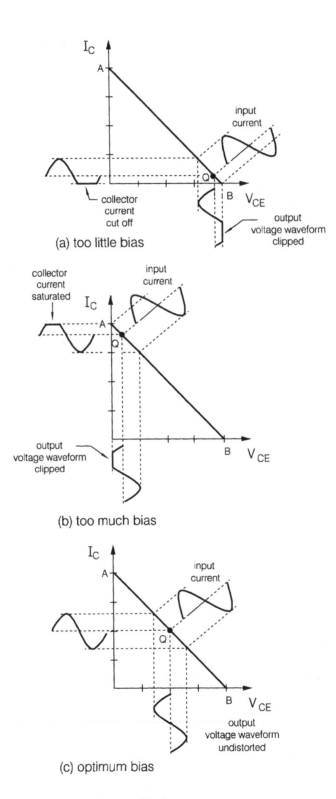

(a) too little bias

(b) too much bias

(c) optimum bias

Figure 3.22 *The importance of correct biasing*

(a) Too little base bias current

Collector current is **cut off** for a part of its negative-going half-cycle, causing the output voltage waveform to be clipped on its **positive** half-cycle.

(b) Too much base bias current

Collector current is **saturated** for a part of its positive half-cycle, causing clipping on the **negative-going** half-cycle of the output voltage waveform.

(c) The optimum bias point

A quiescent point mid-way along the load line enables equal input current swings either side of this point to provide similar equal swings of collector current and output voltage, with the likelihood of minimum distortion of the output signal.

Frequency response

The frequency response characteristic of an amplifier is a measure of the performance of the amplifier over a particular frequency range. This performance can relate to either voltage gain or power gain depending on the nature of the amplifier.

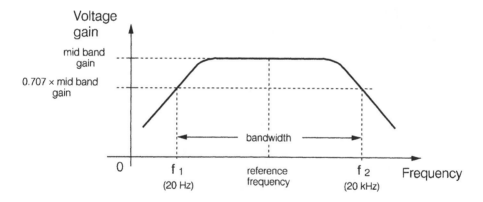

Figure 3.23 *Typical frequency response of audio amplifier*

An example for an audio frequency voltage amplifier is given in Figure 3.23. Ideally, the voltage gain should be constant over the whole of the usable frequency range. The fall-off at both the lower and higher end is due to certain factors which will be dealt with in Practical Exercise 3.5.

Important points

- There are two frequencies, f_1 (low) and f_2 (high), where the voltage gain has fallen to 0.707 of its value at the reference frequency. This reference frequency

---

Continued on p. 74

Important points (*Continued*)

may be either 400 Hz or (more usually) 1 kHz. The range of frequencies between f_1 and f_2, that is, $(f_2 - f_1)$, is known as the **bandwidth** of the amplifier.

- For a power amplifier this bandwidth definition would refer to the two frequencies where the power gain has fallen to $(0.707)^2$, that is, 0.5 of its reference level. These frequencies are sometimes referred to as the **half-power** points.

Practical Exercise 3.5

To obtain the frequency response of a common emitter transistor amplifier. For this exercise you will need the following components and equipment:

1 – npn transistor (BC109)
1 – resistor (470 Ω, 1.5 kΩ, 10 kΩ, 47 kΩ)
3 – capacitor (100 µF, 16 V)
1 – capacitor (0.47 µF, 1000 pF)
1 – DC supply (+12 V)
1 – audio frequency signal generator
1 – Double beam cathode ray oscilloscope
1 – DC voltmeter

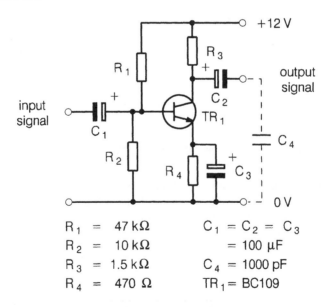

$$R_1 = 47 \text{ k}\Omega \qquad C_1 = C_2 = C_3$$
$$R_2 = 10 \text{ k}\Omega \qquad = 100 \text{ µF}$$
$$R_3 = 1.5 \text{ k}\Omega \qquad C_4 = 1000 \text{ pF}$$
$$R_4 = 470 \text{ }\Omega \qquad TR_1 = BC109$$

Figure 3.24 *The common emitter amplifier: circuit diagram for Practical Exercise 3.5*

Important points

- Figure 3.24 shows the most often-used form of the common emitter amplifier, known as the 'potential divider and emitter resistor' circuit. The full details of its purpose and operation are not within the scope of this book but it will be used for future practical exercises.

- The 1000 pF capacitor is introduced (procedure 8) in order to artificially affect the high frequency response of the amplifier and represents the total capacitance likely to be presented at the transistor output and by the input of a subsequent amplifier stage.

Procedure

1 Connect up the circuit shown in Figure 3.24. The pin connection diagram for the BC109 transistor is shown in Figure 3.5.
2 Calculate the likely DC voltages at the base, emitter and collector with respect to the 0 V line, from

$$V_B = \frac{R_2}{R_1 + R_2} \times V_{CC} \qquad \text{and} \qquad V_E = (V_B - 0.6 \text{ V})$$

$$I_C \approx I_E = \frac{V_E}{R_3} \qquad \text{and} \qquad V_C = V_{CC} - I_C R_L$$

3 Measure these voltages to establish that the DC conditions are correct.
4 Set the input sinusoidal signal frequency to 1 kHz and the voltage to 10 mV peak-to-peak, so that the output voltage is NOT distorted.
5 Measure the output signal voltage and hence calculate the voltage gain.
   This value of gain at 1 kHz will be the **reference**.
   Note. This result and those which follow will be needed in the next chapter.
6 Repeat the above measurements and calculations in order for the voltage gain to be obtained over the frequency range
   (a) 1 kHz to 10 Hz and (b) 1 kHz to 50 kHz.
   Be sure to take measurements at smaller intervals in frequency when the gain changes more rapidly (see Figure 3.23).
7 Plot the frequency response graph.
   At this stage there is unlikely to be a significant fall-off at high frequencies.
8 Add the 1000 pF capacitor C_4 and repeat the measurements of voltage gain from 1 kHz to 50 kHz.
9 Change the value of the coupling capacitors C_1 and C_2 to 0.47 µF and repeat the measurements of voltage gain from 1 kHz to 10 Hz.
10 Add the graphs of the two additional sets of measurements to the previous response.

Continued on p. 76

Practical Exercise 3.5 *(Continued)*

11 Estimate the bandwidth of the amplifier for the conditions $C_1 = C_2 = 0.47\ \mu F$ and $C_4 = 1000$ pF.

   Note. Be sure to **keep these results**.

Important points

- The fall-off at high frequencies is due to the presence of relatively small-value parallel capacitor(s) providing a low impedance path to earth for these frequencies.

- Reducing the value of the coupling capacitors C_1 and C_2 increases the series impedance to the signal, which is more significant at lower frequencies. Figure 3.25 shows the complete picture.

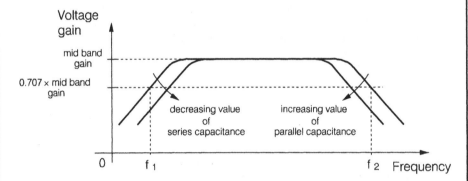

Figure 3.25 *Typical frequency response of an audio amplifier showing effect of additional series and parallel capacitances*

Conclusion

1 For certain amplifiers it is necessary to omit the coupling capacitors altogether. These are known as DC amplifiers, where DC stands for 'direct coupling'. What difference will this make to the frequency response of such an amplifier?

The emitter follower amplifier

This is an amplifier where the output signal is taken from the emitter. It has special properties, which were referred to earlier.

Practical Exercise 3.6

To investigate the action of an emitter follower amplifier.
For this exercise you will need the following components and equipment:

1 – npn transistor (BC109)
1 – resistor (1 kΩ, 10 kΩ, 47 kΩ)
2 – capacitor (100 μF, 16 V)
1 – DC power supply (+12 V)
1 – audio frequency signal generator
1 – double beam cathode ray oscilloscope
1 – DC voltmeter

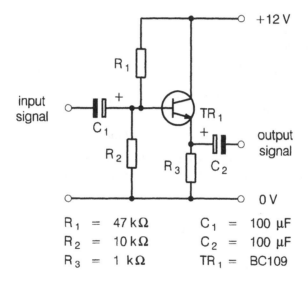

R_1	=	47 kΩ	C_1 =	100 μF
R_2	=	10 kΩ	C_2 =	100 μF
R_3	=	1 kΩ	TR_1 =	BC109

Figure 3.26 *The emitter follower amplifier: circuit diagram for Practical Exercise 3.6*

Procedure

1 Connect up the circuit of Figure 3.26. The pin connection diagram for the BC109 transistor is given in Figure 3.5.
2 Check that the DC conditions are satisfactory by measuring V_B and V_E.
3 Set the signal generator output voltage to 1 V peak-to-peak at 1 kHz.
4 Measure the input and output signals with the CRO and calculate the voltage gain.
5 Note the phase difference between the output and input signal voltages.

Continued on p. 78

Practical Exercise 3.6 (*Continued*)

Conclusion

1 Compare the emitter follower and common emitter amplifiers on the following basis:
  (a) phase shift between output and input signal voltages
  (b) voltage gain.

Important points

Concerning the emitter follower:

- Another name for common collector.

- The load resistor is in the emitter circuit.

- The output signal at the emitter **follows** the input signal at the base – hence the reason for the name **emitter follower**.

- High input resistance and low output resistance.

- Special purpose unit, used for example, for **matching** a high impedance output to a low impedance input, and often, as a **buffer** between units.

The field effect transistor (FET)

Important points

- The name 'field-effect' comes from the fact that the current flowing through an FET is controlled by the **electric field** across its input terminals.

- The FET is a **voltage**-operated device.

- The FET is a **unipolar** device, since current flow consists of **one** type of charge carrier only (holes for p-channel, electrons for n-channel).

- There are various types of FET, namely
    (i) Junction Field Effect Transistor (JFET, JUGFET or simply FET).
    (ii) Metal Oxide Silicon Field Effect Transistor (MOSFET) also called Insulated Gate Field Effect Transistor (IGFET).
    (iii) power FETs such as the **Vertical Metal Oxide Silicon** power field effect transistor (VMOS).

We are concerned here with the JFET, whose circuit symbols are shown in Figure 3.27. The basic arrangement of an n-channel JFET is shown in Figure 3.28. The transistor is made from a bar of n-type material, with non-rectifying contacts at each end for the **drain** and **source**. The two p-type regions are connected together and form the **gate**.

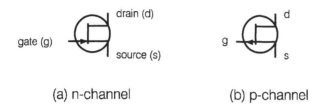

(a) n-channel   (b) p-channel

Figure 3.27 *The junction field effect transistor: circuit symbols*

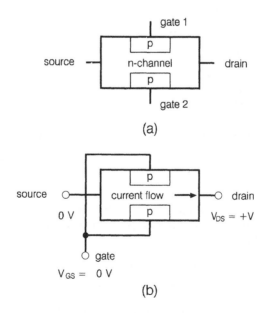

Figure 3.28 *The junction field effect transistor: the basic arrangement*

With the source at 0 V and the gate–source voltage (V_{DS}) initially also at 0 V, a positive voltage between drain and source (V_{DS}) will cause a current flow along the channel from source to drain (I_{DS}).

As the gate–source voltage is made negative, this current will decrease and with sufficient gate voltage, will eventually become zero, that is, cut off. This is illustrated by the transfer characteristic shown in Figure 3.29(a).

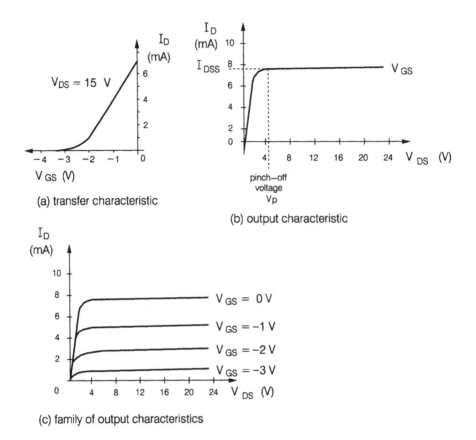

(a) transfer characteristic

(b) output characteristic

(c) family of output characteristics

Figure 3.29   *The junction field effect transistor: transfer and output characteristics*

Important points

- A negative gate–source voltage corresponds to a reverse-biased gate–source pn junction, having little or no gate current, which in turn provides a very **high input resistance** for the FET.

- The drain current is controlled by the voltage across a reverse-biased pn junction.

For a particular value of V_{GS}, increasing V_{DS} causes an increase in I_{DS}. A saturation point for I_{DS} ($= I_{DSS}$) is reached when $V_{DS} = V_P$, the 'pinch-off' point (see Figure 3.29(b)).

A family of curves is obtained for a range of values of V_{GS} (see Figure 3.29(c)).

The FET parameters

The FET is a voltage controlled device in which the drain **current** is controlled by the gate **voltage**. This effect gives a parameter known as **mutual conductance** (g_m) where

$$g_m = \frac{\text{a small change in drain current } I_D}{\text{a small change in gate-source voltage } V_{GS}}$$

for a particular constant value of drain–source voltage V_{DS}.
The shorthand way of writing this is

$$g_m = \left. \frac{\delta I_D}{\delta V_{GS}} \right|_{V_{DS} = \text{(value)} \text{ V}}$$

with the Greek symbol 'δ' (delta) standing for '*a small change in*'.

The units of mutual conductance are current units divided by voltage units which will give the inverse (reciprocal) of ohms, nowadays called siemens (S). The reader may find the units of milliamps per volt (mA/V) in some earlier texts.

An alternative name for mutual conductance is transconductance (g_{fs}) with the same unit of siemens.

Mutual conductance is best found from the transfer characteristic, as follows:

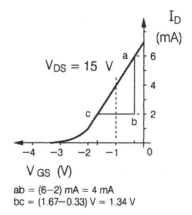

$ab = (6-2) \text{ mA} = 4 \text{ mA}$
$bc = (1.67-0.33) \text{ V} = 1.34 \text{ V}$

Figure 3.30 *The junction field effect transistor: use of transfer characteristic to estimate mutual conductance*

Example 2

Using the transfer characteristic shown in Figure 3.30 estimate the value of the mutual conductance at $V_{GS} = -1$ V.

A tangent is drawn to the graph at the point $V_{GS} = -1$ V and a triangle of reasonable size is formed. The mutual conductance is the gradient or slope of the curve at the chosen point.

The side 'ab' gives us the value for the small change in I_D resulting from the small change in V_{GS} given by side 'bc'.

From the graph, $ab = 4$ mA and $bc = 1.34$ V, giving

$$g_m = \frac{4 \text{ mA}}{1.34 \text{ V}} = 2.98 \text{ mA/V or } 2.98 \text{ mS at } V_{DS} = +15 \text{ V}.$$

[1 milliampere per volt = 1 siemen]

Important point

- The value of g_m will depend upon the value of V_{GS} chosen. For $V_{GS} = -2$ V, g_m would be smaller since the characteristic at this point is less steep.

The other important parameter, which shows the effect that the drain–source **voltage** has on the drain **current**, is known as **drain slope resistance** (r_{ds}) where

$$r_{ds} = \frac{\text{a small change in drain–source voltage } V_{DS}}{\text{a small change in drain current } I_D}$$

for a particular constant value of gate-source voltage V_{GS}.

The shorthand way of writing this is

$$r_{ds} = \frac{\delta V_{DS}}{\delta I_D}\bigg|_{V_{GS}=\text{(value) V}}$$

The units for drain slope resistance are ohms. An alternative name is output conductance (g_{os}) with units of siemens.

Drain slope resistance is most conveniently found from the output characteristic as follows:

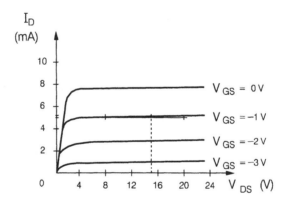

Figure 3.31 *The junction field effect transistor: use of output characteristic to estimate drain slope resistance*

Example 3

Using the output characteristic for $V_{GS} = -1$ V, shown in Figure 3.31, estimate the value of the drain slope resistance when $V_{DS} = 15$ V.

A tangent is drawn to the graph at the point $V_{DS} = 15$ V and a triangle drawn. The drain slope resistance is the **inverse** of the gradient of the curve at the chosen point.

The characteristic is almost horizontal and the formation of the triangle may not be easy. The end result is a very small change in I_D resulting from a small change in V_{DS}.

From the graph,

change in I_D = (5.2 − 5.0) mA = 0.2 mA

change in V_{DS} = (20 − 8) V   = 12 V

Hence,

$$r_{ds} = \frac{12 \text{ V}}{0.2 \text{ mA}} = 60 \text{ k}\Omega \text{ at } V_{GS} = -1 \text{ V}$$

An estimate of drain slope resistance **below** the pinch-off point would have produced a value of about 10 kΩ.

Important points

- For values of V_{DS} below the pinch point, the output slope resistance is low.

- For values of V_{DS} above the pinch point, the output slope resistance is high. The FET is only used in this region.

Practical Exercise 3.7

To obtain the transfer and output characteristics of a junction gate field effect transistor.

For this exercise you will need the following components and equipment:

1 – junction field effect transistor (2N3819)
1 – DC power supply (variable 0 V to +20 V)
1 – DC power supply (variable 0 V to −5 V)
1 – DC ammeter
1 – DC voltmeter

Procedure (a) Transfer characteristic

1 Connect up the circuit shown in Figure 3.32.
2 Set V_{DS} to +15 V and keep this constant.
3 With $V_{GS} = 0$ V, measure I_D.
4 Increase V_{GS} (negatively) in steps and measure I_D, until I_D reaches zero. This should happen at a V_{GS} value of approximately −3.5 V.
5 Complete the table shown in Figure 3.33(a).
6 Plot the graph of I_D (vertically) against V_{GS} in the appropriate quadrant. Refer to Figure 3.29(a).
7 Estimate the value of mutual conductance (g_m) at $V_{GS} = -1$ V.

Continued on p. 84

Practical Exercise 3.7 (*Continued*)

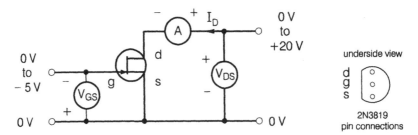

Figure 3.32   *The junction field effect transistor characteristics: circuit diagram for Practical Exercise 3.7*

Procedure (b) Output characteristics

See Figure 3.32.

1 Set V_{GS} to 0 V and keep this constant.
2 Vary V_{DS} from 0 V to +20 V in steps and measure I_{DS}.
3 Repeat procedure 2 for V_{GS} values of −1.0 V, −2.0 V, and −3.0 V.
4 Complete the table shown in Figure 3.33(b).
5 Plot the graph of I_D (vertically) against V_{DS}, for each value of V_{GS}, to give the family of characteristics.

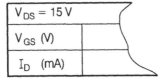

$V_{DS} = 15$ V	
V_{GS} (V)	
I_D (mA)	

(a) transfer characteristic

V_{GS} (0 V)		V_{GS} (−1 V)		V_{GS} (−2 V)		V_{GS} (−3 V)	
V_{DS} (V)	I_D (mA)	V_{DS} (V)	I_D (mA)	V_{DS} (V)	I_D (mA)	V_{DS} (V)	I_D (mA)

(b) output characteristics

Figure 3.33   *Table of results for Practical Exercise 3.7*

---

6 Estimate the value of drain slope resistance (r_{ds}) on the curve for $V_{GS} = -1$ V, for a value of V_{DS}

(a) below the pinch-off point ($V_{DS} = +2$ V)
(b) above the pinch-off point ($V_{DS} = +15$ V).

The JFET amplifier

As with the bipolar transistor, there are three modes of operation, namely

(i) common source (corresponding to common emitter)
(ii) common gate (common base)
(iii) common drain (common collector).

The basic circuit of a common source amplifier is shown in Figure 3.34(a).

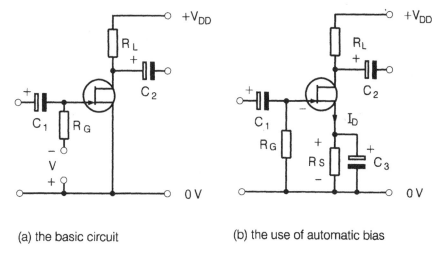

(a) the basic circuit   (b) the use of automatic bias

Figure 3.34 *The common source amplifier*

The input signal voltage is supplied to the gate–source circuit through the coupling capacitor C_1. The output signal voltage is developed across the drain load resistor R_L as the result of the current through it and coupled through capacitor C_2 to the outside world.

In practice, the separate DC supply for the gate–source voltage is provided by **automatic biasing** shown by the circuit in Figure 3.34(b). The method of deriving this bias is as follows:

(a) The drain current (DC) flows through R_S giving a voltage (DC) of $I_D R_S$ across R_S, with a polarity $\pm$ as shown.

(b) The value of R_G will be large (MΩ), there will be negligible gate current and thus no current flows through R_G.

(c) No voltage will be developed across R_G and so the voltage at the gate with respect to the source, and its polarity, will be the **same** as that at the lower end of R_S.

(d) The gate voltage is **negative** with respect to the source and equal to $I_D R_S$.

Capacitor C_3 acts as a bypass for the AC signal. This will be referred to again in the chapter dealing with negative feedback.

Practical Exercise 3.8

To investigate the action of a common source field effect transistor amplifier.
For this exercise you will need the following components and equipment:

1 – DC supply (+15 V)
1 – junction field effect transistor (2N3819)
1 – resistor (4.7 kΩ, 33 kΩ, 1 MΩ)
2 – capacitor (0.1 μF)
1 – capacitor (100 μF, 25 V)
1 – audio frequency signal generator
1 – double beam cathode ray oscilloscope
1 – DC voltmeter

Procedure

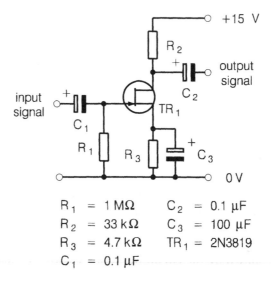

R_1 = 1 MΩ C_2 = 0.1 μF
R_2 = 33 kΩ C_3 = 100 μF
R_3 = 4.7 kΩ TR_1 = 2N3819
C_1 = 0.1 μF

Figure 3.35   *The common source amplifier: circuit diagram for Practical Exercise 3.8*

1 Make up the circuit shown in Figure 3.35. The pin connection diagram for the 2N3819 transistor is shown in Figure 3.32.

2 Check the **DC operating conditions** by measuring the voltage with respect to the 0 V line:

(a) across R_S (V_{RS}). This should be approximately 1 V.

(b) at the drain (V_D). This should be approximately 8 V.

3 The drain–source voltage (V_{DS}) can be calculated as

$$V_{DS} = V_D - V_{RS}$$

which should then be approximately one-half of the supply voltage.

Important points

- The figure for V_{RS} gives the value of the bias voltage, V_{GS}, with the gate being negative with respect to the source.

- A direct measurement of the voltage between gate and source could give a different answer depending on the type of meter used. Why is this?

---

4 Set the frequency of the signal generator to 1 kHz and the output level to 100 mV peak-to-peak.

5 Measure the output voltage and hence calculate the voltage gain of the amplifier.

6 Using the value for g_m and r_{ds} (high) obtained in Practical Exercise 3.7, calculate the theoretical value of the voltage gain.

The load line

Refer to Figure 3.36. The technique for establishing the position of the DC load line resembles that given earlier for the npn transistor.

The resistor R_3 must be included as follows:

(a) The DC condition is given by

$$V_{DD} = I_D R_L + V_{DS} + V_{RS}$$

$$= I_D R_L + V_{DS} + I_D R_S$$

$$= I_D (R_L + R_S) + V_{DS}$$

(b) Point A on the graph is found when $V_{DS} = 0$ V, giving

$$I_D = \frac{V_{DD}}{R_L + R_S}$$

(c) Point B on the graph is found when $I_D = 0$ mA, giving

$$V_{DS} = V_{DD}$$

(d) Joining the points A and B gives the DC load line for the particular value of drain load resistor R_L and supply voltage V_{DD}.

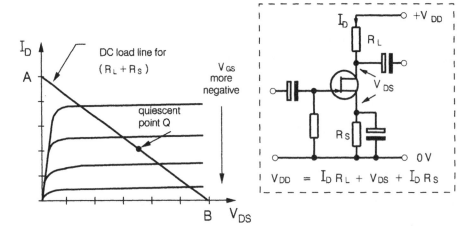

Figure 3.36 *The common source amplifier load line: derivation*

Important points

- The intersection of the DC load line with the required value of gate–source bias voltage curve gives the quiescent point (Q).

- The values of I_D and V_{DS} where lines from the Q point intersect the respective axes are the no-signal values.

- The application of an input signal voltage causes the gate–source voltage to vary above and below the bias value.

- A change in gate–source voltage causes a corresponding change in drain–source voltage with the result that the voltage gain is given as

$$\text{voltage gain} = \frac{\text{change in drain–source voltage}}{\text{change in gate–source voltage}}$$

Amplifier models

The performance of a system can be analysed by replacing the system with its equivalent circuit, this being an electrical network in terms of the parameters of the system. A typical model outline is the four-terminal network, shown in Figure 3.37.

The bipolar transistor is most conveniently represented by a set of so-called hybrid or h-parameters, as follows:

Input resistance $\qquad\qquad (h_i) = \dfrac{V_i}{I_i}$, with V_o zero

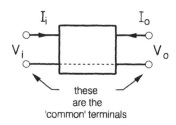

Figure 3.37 *The four-terminal network*

	The unit is the ohm (Ω)
Current gain	$(h_f) = \dfrac{I_o}{I_i}$, with V_o zero
	No units, just a ratio
Output conductance	$(h_o) = \dfrac{I_o}{V_o}$, with I_i zero
	The unit is the siemen (S)
Reverse voltage ratio	$(h_r) = \dfrac{V_1}{V_2}$, with I_i zero
	No units, just a ratio

Important points

- The parameters are named 'hybrid' because their dimensions are **mixed** (a definition of hybrid).

- The statement 'V_o zero' means that the parameters are measured with the output **short-circuited** to **AC** signals.

- The statement 'I_i zero' means that the parameters are measured with the input **open-circuited** to **AC** signals.

- For a particular configuration the appropriate additional subscript is added. Thus, for common emitter, the parameters are h_{ie}, h_{fe}, h_{oe} and h_{re}.

- The use of small letters for the subscripts means that these parameters refer to the AC circuit or 'signal' conditions.

Hybrid parameter values

The table in Figure 3.38 shows a typical set of h-parameter values for a common emitter BC109 transistor.

$$h_{ie} = 8.8\,k\Omega$$
$$h_{oe} = 60\,\mu S$$
$$h_{re} = 3 \times 10^{-4}$$
$$h_{fe} = 600$$

Figure 3.38   *The BC109 transistor: common emitter hybrid parameters*

The common emitter hybrid equivalent circuit

The starting point is the circuit in Figure 3.39(a).

An explanation of the less obvious components in this circuit should help.

(a) $h_r V_o$ is a voltage fed back from the output circuit and shown as a voltage generator in series with h_{ie}.

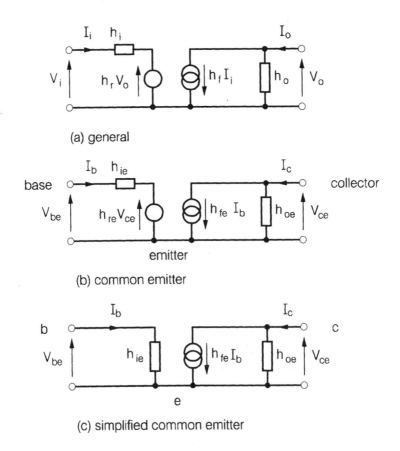

(a) general

(b) common emitter

(c) simplified common emitter

Figure 3.39   *The bipolar transistor equivalent circuit using hybrid parameters*

(b) $h_f I_i$ is a current generator, being the current gain (h_f) times the input current (I_i).

The common emitter version is shown in Figure 3.39(b).

As usual, any justifiable simplification we can make will help the cause! In practice, the value of h_{re} is sufficiently small to enable the voltage generator $h_{re}V_{ce}$ to be omitted from the circuit with no appreciable inaccuracy. The simplified circuit is shown in Figure 3.39(c).

We will now do a problem, while bearing in mind the following:

Important points

• Coupling capacitors may be regarded as short-circuits to the AC signal, and therefore need not be included, together with any resistors which themselves are short circuited by capacitors.

• The DC supply is a very low resistance path to the signal frequencies and can be replaced (in the equivalent circuit only!) by a short circuit.

Example 4

The transistor used as an amplifier in the circuit shown in Figure 3.40 has the h-parameters given in Figure 3.38.

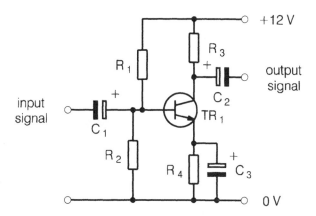

Figure 3.40 *Circuit diagram for Example 4*

(i) Draw the equivalent circuit.

This is shown in Figure 3.41.

(ii) Calculate (a) the effective input resistance, (b) the output resistance, (c) the current gain and (d) the voltage gain, if component values are as follows:

$R_1 = 47 \text{ k}\Omega$, $R_2 = 10 \text{ k}\Omega$ and $R_L = 1.5 \text{ k}\Omega$

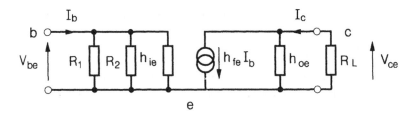

Figure 3.41   *The bipolar transistor equivalent circuit for Example 4*

(a) The effective input resistance is the result of R_1, R_2 and h_{ie} in parallel.

$$h_{ie} = 8.8 \text{ k}\Omega$$

Thus from

$$\frac{1}{R_{in}} = \frac{1}{R_1} + \frac{1}{R_2} + \frac{1}{h_{ie}}$$

$$R_{in} = 4.3 \text{ k}\Omega$$

The larger the values of R_1 and R_2, the closer R_{in} will be to the value of h_{ie}.

(b) The output resistance is the reciprocal of h_{oe}.

$$h_{oe} = 60 \text{ μS}$$

Thus

$$R_{out} = \frac{1}{h_{oe}}$$

which gives $R_{out} = 16.7 \text{ k}\Omega$

(c) The current gain (A_i) is given by $\dfrac{I_c}{I_b}$

$$I_c = -\frac{V_{ce}}{R_L} \quad \text{and} \quad V_{ce} = -h_{fe}I_b R'_L$$

where R'_L is the result of R_L and $1/h_{oe}$ in parallel.
The 'minus' signs here are the result of the assumed direction of the output current.
Thus,

$$I_c = \frac{h_{fe}I_b R'_L}{R_L}$$

and current gain (A_i) $= \dfrac{I_c}{I_b} = \dfrac{h_{fe}R'_L}{R_L}$

$$R_L = 1.5 \text{ k}\Omega \quad \text{and} \quad \frac{1}{h_{oe}} = R_{out} = 16.7 \text{ k}\Omega,$$

giving $R'_L = 1.38 \text{ k}\Omega$.
Then,

$$\text{current gain} = \frac{600 \times 1.38 \text{ k}\Omega}{1.5 \text{ k}\Omega} = 552$$

(d) The voltage gain (A_v) is given by V_{ce}/V_{be}

$$V_{ce} = -h_{fe}I_bR'_L \quad \text{and} \quad V_{be} = I_bR_{in}$$

$$R_{in} = 4.3 \text{ k}\Omega,$$

thus voltage gain

$$(A_v) = \frac{V_{ce}}{V_{be}} = -\frac{h_{fe}R'_L}{R_{in}}$$

$$A_v = -\frac{600 \times 1.38 \text{ k}\Omega}{4.3 \text{ k}\Omega} = -192$$

The 'minus' sign represents the 180° phase shift between output and input signal voltages for the common emitter amplifier.

Important points

- If the effect of h_{oe} is negligible, which is sometimes the case, then R_L becomes R_L, resulting in the current and voltage gains being as follows:

Current gain $(A_i) = h_{fe}$

Voltage gain $(A_v) = \frac{h_{fe}R_L}{R_{in}}$

- Furthermore, if $R_{in} = h_{ie}$, then

Voltage gain $(A_v) = \frac{h_{fe}R_L}{h_{ie}}$

Questions

3.7 Figure 3.42 shows the small-signal equivalent circuit of a common emitter transistor amplifier with the corresponding hybrid parameters. The load resistance is 4.7 kΩ.

Calculate (a) the input resistance, (b) the current gain and (c) the voltage gain.

3.8 Draw the hybrid equivalent circuit of the amplifier shown in Figure 3.43. If $h_{ie} = 5$ kΩ, $h_{fe} = 200$, $h_{oe} = 15$ μS and h_{re} is negligible, determine:

(a) the effective input resistance
(b) the output resistance
(c) the current gain
(d) the voltage gain.

Continued on p. 94

Questions (*Continued*)

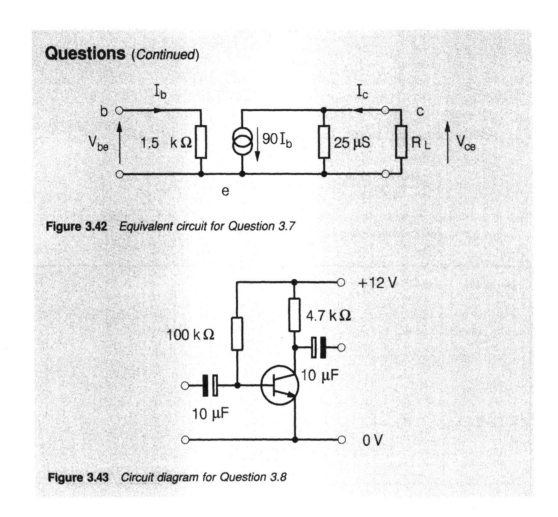

Figure 3.42 *Equivalent circuit for Question 3.7*

Figure 3.43 *Circuit diagram for Question 3.8*

The field effect transistor equivalent circuit

The common source equivalent circuit is shown in Figure 3.44.

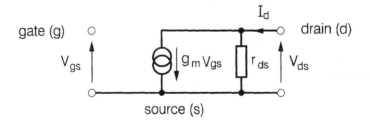

Figure 3.44 *The field effect transistor equivalent circuit*

Important points

- An open-circuit is shown between gate and source because, under normal operating conditions, the gate current is virtually zero.

- The FET can be replaced by a constant current generator $g_m V_{gs}$ in parallel with its drain–slope resistance r_{ds}.

- Current gain for the FET, while being extremely large in value, has no real meaning.

The equivalent circuit will now be used to solve a problem.

Example 5

The circuit of a field effect transistor amplifier is shown in Figure 3.45. Draw the equivalent circuit and from it determine the value of (a) the input resistance and (b) the voltage gain, if $r_{ds} = 60$ kΩ and $g_m = 3$ mS $= 3$ mA/V and $R_L = 33$ kΩ.

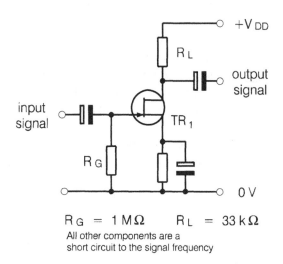

$$R_G = 1\,M\Omega \qquad R_L = 33\,k\Omega$$

All other components are a
short circuit to the signal frequency

Figure 3.45 *Circuit diagram for Example 5*

(a) The effective input resistance (R_{in}) is the parallel combination of R_G and the input resistance of the FET itself. Since the latter is very high (many MΩ), then R_{in} will be equal to R_G.

(b) The voltage gain (A_v) is given by V_{ds}/V_{gs}.
$V_{ds} = g_m V_{gs} R'_L$, where R'_L is the parallel combination of R_L and r_{ds}.
That is,

$$R'_L = \frac{r_{ds} R_L}{r_{ds} + R_L}$$

Thus,

$$\text{voltage gain } (A_v) = \frac{V_{ds}}{V_{gs}} = g_m R'_L$$

which gives $A_v = \dfrac{g_m r_{ds} R_L}{r_{ds} + R_L}$

$r_{ds} = 60 \text{ k}\Omega, g_m = 3 \text{ mS and } R_L = 33 \text{ k}\Omega.$

$$\text{Therefore } A_v = \frac{3 \times 10^{-3} \times 60 \times 10^{+3} \times 33 \times 10^{+3}}{93 \times 10^{+3}} = 63.9$$

Important point

- If r_{ds} is very much greater than R_L, then $R'_L = R_L$, thus giving $A_v = g_m R_L$.

Questions

3.9 In a common source amplifier the FET parameters are $g_m = 2 \text{ mS}$ and $r_{ds} = 100 \text{ k}\Omega$.

(a) If the load resistance is 47 kΩ, calculate the voltage gain of the amplifier.

(b) What value of load resistor is required to give a voltage gain of 20?

3.10 A certain FET has a mutual conductance of 2.5 mS and is used as a common source amplifier with a drain load of 10 kΩ. Calculate the voltage gain of the amplifier assuming that the drain–slope resistance is

(a) 100 kΩ

(b) very much larger than the drain load.

3.11 Consult the catalogue data to find the value of g_m for a range of n-channel FETs.

Hint You may need to look for these under the symbol of Y_{fs} which stands for 'forward transfer admittance'!

Practical Exercise 3.9

To investigate the action of a source follower amplifier.
For this exercise you will need the following components and equipment:

1 – field effect transistor (2N3819)
1 – resistor (1 MΩ)
1 – resistor (1 kΩ)
2 – capacitor (100 µF, 25 V)
1 – DC power supply (+15 V)

1 – audio frequency signal generator
1 – double beam cathode ray oscilloscope

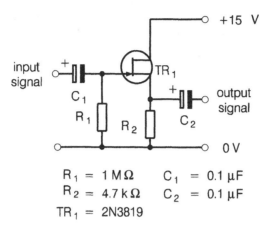

$$R_1 = 1\,M\Omega \qquad C_1 = 0.1\,\mu F$$
$$R_2 = 4.7\,k\Omega \qquad C_2 = 0.1\,\mu F$$
$$TR_1 = 2N3819$$

Figure 3.46   *The source follower amplifier: circuit diagram for Practical Exercise 3.9*

Procedure

1 Make up the circuit of Figure 3.46. The pin connection diagram for the 2N3819 transistor is given in Figure 3.32.
2 Measure the DC voltage between the source and earth. This will be the gate–source bias voltage.
3 Set the signal generator output voltage to 1 V peak-to-peak at 1 kHz.
4 Measure the input and output signals with the CRO and calculate the voltage gain.
5 Note the phase difference between the output and input signal voltages.

Conclusion

1 Compare the source follower and common source amplifiers on the following basis:
   (a) phase shift between output and input signal voltages
   (b) voltage gain.

Comparison of the bipolar and FET unipolar transistors

The silicon bipolar transistor has good voltage and power gains, combined with low cost. Since it is current controlled, it has a low-to-medium input impedance. Because of these features, it finds many suitable applications.

The FET on the other hand is voltage controlled, giving it a very high input impedance. This, combined with its low noise, makes it preferred for input stages of audio frequency amplifiers. Other features of the FET include good linearity, giving low distortion levels, and a good high frequency performance because of its low interelectrode capacitance, essential for radio frequency application. It is generally regarded as a good switching device.

4 Logarithmic units

The human ear response

The human ear is considerably sensitive to changes in the pitch or frequency of a sound, but less so to changes in amplitude or volume. The ear responds to proportional changes in sound levels rather than to the absolute levels themselves. Put another way, the response is said to be **logarithmic**. This means that the impression a listener receives when a sound is increased in volume is proportional to the **ratio** of the energy or power of the two sound levels rather than to the absolute increase in levels. For example, for every doubling of sound intensity, an equal change in loudness is heard.

Example 1

The power output from a loudspeaker increases from 1 W to 10 W, which represents an absolute increase of 9 W and an output to input ratio of 10 to 1.

If the power is increased from 10 to 100 W, the increase is now 90 W with the **ratio** of output to input **unchanged** at 10 to 1.

The human ear recognizes these two increases in loudness as being equal, in spite of the fact that the absolute increases in actual power levels are very different.

One definition of a logarithmic scale is 'a scale of measurement in which an increase of one unit represents a tenfold increase in the quantity measured'. To help the understanding of this statement, let's look at Figure 4.1, which shows the axes that are used to draw a frequency response graph.

Important points

- The vertical scale is **linear** since it shows equal increases in voltage gain for equal increases in distance.

- The horizontal scale is **logarithmic** since it shows equal **ratios** of frequency increase for equal increases in distance.

From the mathematics, one type of logarithmic unit (called common logarithms) is concerned with powers of 10.

We know that (a) $10^1 = 10$

(b) $10^2 = 100$

(c) $10^3 = 1000$ and so on

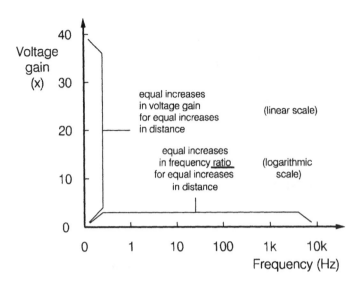

Figure 4.1 *Frequency response graph: linear and logarithmic scales*

The mathematics tell us that the logarithm (to the base of 10) of

(a) 10 is 1, and is written as $\log_{10} 10$ $= 1$
(b) 100 is 2, and is written as $\log_{10} 100$ $= 2$
(c) 1000 is 3, and is written as $\log_{10} 1000 = 3$ and so on.

The need for 'tables of logarithms' has long since gone, since the ubiquitous scientific calculator can do it on our behalf. For those that would appreciate a memory refresher, please do the following examples on your calculator:

Example 2

(i) To find $\log_{10} 10$:
- enter '10' into the display
- press 'log' (this may have '10^x' written above it)
- the display will give the answer '1' as being the **logarithm of 10** to the base of 10.

(ii) To find $\log_{10} 100$:
- enter '100' into the display
- press 'log'
- the display will give the answer '2' as being the **logarithm of 100** to the base of 10.

The process works in reverse as follows:

(iii) To find the number whose logarithm to the base of 10 is 1:
- enter '1' into the display

- press 'shift' then 'log' (this is known as 'antilog')
- the display will give the answer '10' as being the **antilogarithm of 1** to the base of 10.

Amplifier gain

With amplifiers the expectation is that the output signal will be larger than the input signal. In any case, the expression for gain is given by

$$\text{Gain} = \frac{\text{signal out}}{\text{signal in}}$$

Example 3

(i) The input signal is 1 V and the output signal 5 V.
   The amplifier voltage gain (A_v) is given by

$$A_v = \frac{5\ V}{1\ V} = 5\ \text{(no units)}$$

   We say that the output voltage is 5 **times** the input voltage.

(ii) The power input and outputs for a certain amplifier are 1 milliwatt (1 mW) and 1 W respectively.
   The amplifier power gain A_P is given by

$$A_P = \frac{1\ W}{1\ mW} = 1000\ \text{(again, no units)}$$

   The output power is 1000 **times** the input power.

For examples where the output signal is smaller than the input, the term amplifier may be thought to be a misnomer. For a moment let's continue with that name.

(iii) The input signal is 1 V and the output signal is 0.1 V.
   The amplifier voltage gain A_v is given by

$$A_v = \frac{0.1\ V}{1\ V} = 0.1\ \text{(no units)}$$

   The opposite to gain is loss, otherwise known as **attenuation**. In this example we say that the attenuation is 10 (times), namely the inverse of 0.1.
   Let us now move away from linear to logarithmic units.

The decibel

The decibel is a way of expressing amplifier gains and losses as ratios. By definition

$$\text{power gain} = \log_{10} \frac{\text{output power } (P_o)}{\text{input power } (P_i)}$$

The unit is the bel (B), named after Alexander Graham Bell.

The bel is in fact a large unit and the practical unit is the decibel (dB), where the decibel is one-tenth of a bel. Thus

$$\text{power gain} = 10 \log_{10} \frac{\text{output power } (P_o)}{\text{input power } (P_i)} \text{ (dB)}$$

Example 4

(i) An amplifier has a power gain of 100 times.

$$\text{Power gain} = 10 \log_{10} 100$$

$$= 10 \times 2$$

$$= 20 \text{ dB}$$

For convenience, the base (of 10) will no longer be written and the word 'log' will be taken to mean '$\log_{10}$'.

(ii) A network has a power gain of 0.5 times (alternatively, an attenuation of 2 times).

$$\text{Power gain} = 10 \log \frac{\text{output power } (P_o)}{\text{input power } (P_i)} \text{ (dB)}$$

$$= 10 \log 0.5$$

$$= 10 \times (-0.301)$$

$$= -3.01 \text{ dB}$$

Both of the following statements are correct:

'the power gain is −3.01 dB'

'the attenuation is 3.01 dB'

This value of 3.01 is important, since it gives the decibel equivalent of half-power. It will be used later, as a rounded-down figure of 3 dB.

(iii) A certain amplifier has a gain of 40 decibels. What exactly does this mean? Can we work out what the actual input and output powers are?

Compare this with 'car A costs 3 times as much as car B'. Just how much do each of these cars actually cost?

The answer, of course, is that it is impossible to say, without having a vital piece of information called a **reference level**.

In these examples this information might be the cost of either car A or car B, or the input or output power of the amplifier.

For the amplifier, suppose the input power was 10 mW. Since the power gain is 40 dB, the output power level is **40 dB above 10 mW**.

From

$$\text{power gain} = 10 \log \frac{\text{output power } (P_o)}{\text{input power } (P_i)} \text{ (dB), we can write}$$

$$40 = 10 \log \frac{P_o}{10 \times 10^{-3}}$$

(using the basic units of watts)

$$4 = \log \frac{P_o}{10 \times 10^{-3}}$$

$$\frac{P_o}{10 \times 10^{-3}} = \text{antilog } 4 = 10\,000$$

giving $P_o = 10 \times 10^{-3} \times 10\,000$

$$= 10 \text{ W}$$

Thus the output power is calculated from a knowledge of the input.

The reference level

It is general practice to specify a reference level that is intended to be neither the input nor the output of a particular system. The commonly used reference level for electronics and telecommunications is **1 mW** and decibel levels relative to this are given the notation of **dBm**, with 1 mW being 0 dBm.

Important points

- The decibel is a **ratio** and not a level of absolute power.

- A reference level is needed in order to provide **actual** input and output levels.

Advantages in using decibel units

Consider the following:

Example 5

A number of amplifiers are connected in series, with respective power gains as shown in Figure 4.2.

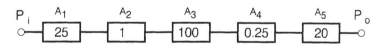

Figure 4.2 *Diagram for Example 5*

The overall linear power gain A_p is found from

$$A_p = \frac{P_o}{P_i} = A_1 \times A_2 \times A_3 \times A_4 \times A_5$$

$$= 25 \times 1 \times 100 \times 1/4 \times 40$$

$$= 25\,000$$

In decibel units,

$$A_p = 10 \log 25\,000$$

$$= 10 \times 4.39$$

$$= 43.9 \text{ dB}$$

The mathematical rules are that where linear units are multiplied together, their logarithmic equivalents are added. Thus we see that by converting each linear gain into its respective logarithmic unit, the overall power gain in dB can be found from

$$A_p = (13.97 + 0 + 20 - 6.02 + 16.02) \text{ dB}$$

$$= 43.9 \text{ dB}$$

The advantages in using logarithmic rather than linear units are summarized:

Important points

- Logarithmic units will generally provide smaller and less cumbersome numbers, since they are added algebraically instead of being multiplied.

- The logarithmic response of the human ear makes for easier presentation of information, with frequency axes scaled in ratios.

- Volume controls (but not tone controls) for audio equipment obey a 'logarithmic law'.

Voltage and current ratios

Although the decibel **must be** defined in terms of power ratios, it is frequently necessary to express voltage ratios (and sometimes current) in a similar manner.

To find the dB gain or loss from a voltage ratio, the procedure is as follows:

The power P developed in a resistor R as a result of a voltage V across the resistor can be found from

$$P = \frac{V^2}{R}$$

Thus the power ratio P_o/P_i provides the expression

$$\frac{P_o}{P_i} = \frac{V_o^2/R_o}{V_i^2/R_i}$$

$$= \frac{V_o^2 R_i}{V_i^2 R_o}$$

Now, if and **only if** $R_i = R_o$, this simplifies to

$$\frac{P_o}{P_i} = \frac{V_o^2}{V_i^2}$$

$$= \left(\frac{V_o}{V_i}\right)^2$$

The decibel gain resulting from this is given by

$$10\log\frac{P_o}{P_i} = 10\log\left(\frac{V_o}{V_i}\right)^2 = 20\log\frac{V_o}{V_i} \quad \text{(using the rule of logs for indices)}$$

Similarly for current ratios,

$$\text{dB gain} = 20\log\frac{I_o}{I_i}$$

This expression is used less often than that for voltage ratios.

The respective expressions using voltage and current ratios will provide a value of dB gain identical to the one using power ratios, **providing the input and output resistances are equal**. This proviso is rarely met in practice (transmission lines excepted) but in spite of this, the expression is widely used without consideration of this need. The result is that the gains using power and voltage (or current) ratios will no longer be identical. To understand this point, consider the following example, arising from the situation shown in Figure 4.3.

Example 6

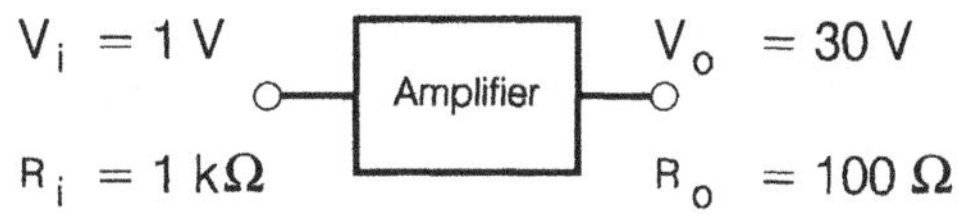

$$V_i = 1\text{ V} \qquad \boxed{\text{Amplifier}} \qquad V_o = 30\text{ V}$$
$$R_i = 1\text{ k}\Omega \qquad\qquad\qquad R_o = 100\ \Omega$$

Figure 4.3 *Diagram for Example 6*

$$\text{voltage gain} = \frac{30}{1} = 30 \text{ times, and in dB,}$$

$$\text{voltage gain} = 20\log 30$$

$$= 20 \times 1.5$$

$$= 30\text{ dB}$$

$$\text{input power} = \frac{V^2}{R}$$

$$= \frac{1^2}{1000}$$

$$= 1\text{ mW}$$

$$\text{output power} = \frac{30^2}{100} = 9\text{ W}$$

Hence,

$$\text{power gain} = 10 \log \frac{9}{1 \times 10^{-3}}$$

$$= 10 \log 9000$$

$$= 39.5 \text{ dB}$$

If R_i and R_o had been equal in value, for example say 1000 Ω, then with $V_{in} = 1$ V
and $V_{out} = 30$ V
As before,

voltage gain $= 30$ dB and

input power $= 1$ mW

$$\text{output power} = \frac{30^2}{1000}$$

$$= 0.9 \text{ W, giving}$$

$$\text{power gain} = 10 \log \frac{0.9}{1 \times 10^{-3}}$$

$$= 10 \log 900$$

$$= 30 \text{ dB}$$

$$= \text{voltage gain}$$

There are several important decibel values, related to respective power and voltage gains, which are shown in Figure 4.4. These values should be confirmed by calculation and committed to memory.

power gain $\times 2$, dB $= 10 \log 2 = 10 \times 0.3 = 3$

power gain $\times 0.5$, dB $= 10 \log 0.5 = 10 \times -0.3 = -3$

dB gain	Power gain	Voltage gain
0	1	1
3	2	1.414
6	4	2
10	10	3.16
20	100	10
−10	0.1	0.316
− 6	0.25	0.5
− 3	0.5	0.707

Figure 4.4 *Table giving important decibel values*

voltage gain × 2,   dB = 20 log 2 = 20 × 0.3 = 6

voltage gain × 0.5,   dB = 20 log 0.5 = 20 × −0.6 = −6

Example 7

The gain of an amplifier at a certain frequency falls by 3 dB. What does this mean for changes in voltage and power?

A logarithmic change of −3 dB = a linear power change of 0.5 (see above)

= a linear voltage change of 0.707

The meaning is therefore that a fall of 3 dB represents

(a) a change in power output of one-half of its value at the reference level of 0 dB
(b) a change in voltage output of 0.707 of its reference value

These are very important values and should also be remembered. An important practical application for these values occurs with regard to frequency response specifications and measurements.

It would be helpful now to refer to the frequency response graph diagram drawn as a result of Practical Exercise 3.5. The exercise will now be extended to include work on decibels.

Practical Exercise 4.1

To plot amplifier frequency response using logarithmic units.
  No components or equipment required.

Procedure

1 Refer to the results obtained earlier in Practical Exercise 3.5 for amplifier frequency response and convert the values of voltage gain into decibels.
2 Plot the graph of dB voltage gain (vertically) against frequency. A typical graph is shown in Figure 4.5 where this vertical scale is on the left-hand side.

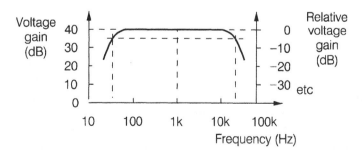

Figure 4.5 *Amplifier frequency response: typical graph for Practical Exercise 4.1*

Continued on p. 108

Practical Exercise 4.1 *(Continued)*

A more convenient way of presenting this information is by giving the gain at the reference frequency a value of 0 dB. All other gains will then be relative to this figure of 0 dB. This is shown by the scale on the right-hand side of the graph.

3 Add a second vertical axis and put on the values using the gain at 1 kHz as the reference 0 dB.

Important points

- The frequencies f_1 and f_2 where the gain has fallen to 3 dB below its value at the reference frequency are used to determine the **bandwidth** of the amplifier.
- Bandwidth $= f_2 - f_1$.

---

4 Determine the bandwidth of this amplifier.

Conclusions

1 Two separate 50-W amplifier specifications read:
  (a) 'frequency response (−3 dB) 20 Hz to 20 kHz'
  (b) 'frequency response (−1 dB) 20 Hz to 20 kHz'

  Which is the 'better' amplifier over this frequency range, from the power output point of view, and why?
2 Obtain a manufacturer's leaflet for an audio amplifier and examine the specification. In particular, look for:

  (a) rms power output (continuous sine wave) for a given load
  (b) frequency response at ±1 dB and ±3 dB.

Decibels and sound levels

Reference was made earlier to the human ear and ratios of sound levels. One use of the decibel unit is indeed to express sound intensities. Again, it is important to remember that for this, the decibel is a ratio of loud-to-soft and not an actual sound level. An investigation very many years ago provided the fact that a large orchestra playing at its loudest created the acoustic equivalent of 70 electrical watts. A violin by contrast, at its quietest, produced 3.8 microwatts (1 micro $= 1 \times 10^{-6}$). The result is a power ratio of 18 million to 1, or 72 dB, which equates to a sound pressure ratio of 4250 to 1. The value of 72 dB is known as the dynamic range of the orchestra.

For everyday practical use, the reference sound level of 0 dBA, called the threshold of hearing, is the sound which can just be heard by the human ear. The other end of the scale, the threshold of pain, is where the sound level is such as to cause physical discomfort and is generally regarded as being 120 dBA. The 'A' refers to a weighting process in which the frequency response matches that of the human ear. The smallest change in audio power that can be appreciated by the human ear is said to be 1 dBA. A list of sounds with their decibel value is given in Figure 4.6.

Sound	dB level
Threshold of hearing	0
Whisper	30
Conversation	60
Busy street	80
Threshold of pain	120
Jet engine	140

Figure 4.6 *Typical sound levels in decibels*

Practical Exercise 4.2

To investigate the pi and tee attenuator networks.
For this exercise you will need the following components and equipment:

1 – resistor (56 Ω, 75 Ω, 100 Ω)
2 – resistor (12 Ω, 27 Ω, 220 Ω, 470 Ω)

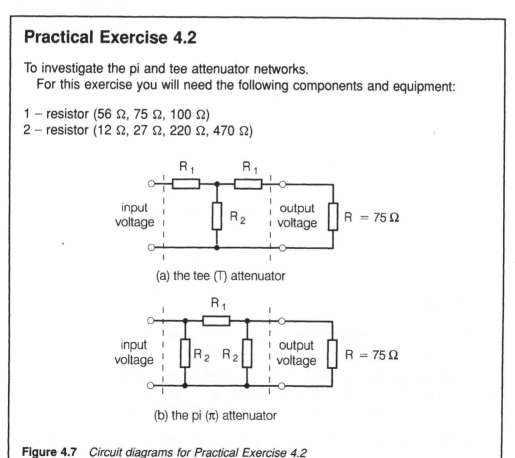

(a) the tee (T) attenuator

(b) the pi (π) attenuator

Figure 4.7 *Circuit diagrams for Practical Exercise 4.2*

Continued on p. 110

Practical Exercise 4.2 (*Continued*)

1 – audio frequency signal generator
1 – double beam cathode ray oscilloscope (or electronic voltmeter)

Procedure

Note. The attenuator network **must** be terminated in a resistor having a value equal to the characteristic resistance of the network, in this example, 75 Ω.
 Using each of the circuits of Figure 4.7 in turn:

1 Measure the output voltage for an input voltage of 1 V (rms or peak-to-peak).
2 Calculate the ratio V_o/V_i and express this in dB.
3 Complete the table in Figure 4.8.

Network	R_1 (Ω)	R_2 (Ω)	Attenuation Measured	Calculated
(tee)	12	220	$\frac{V_o}{V_i}=$	dB
	27	100	$\frac{V_o}{V_i}=$	dB
(pi)	27	470	$\frac{V_o}{V_i}=$	dB
	56	220	$\frac{V_o}{V_i}=$	dB

Figure 4.8 *Table of results of Practical Exercise 4.2*

Questions

4.1 Express as decibel gains or losses the following power ratios:
 (a) 1000 (b) 500 (c) 250 (d) 1 (e) 0.1 (f) 0.01
4.2 The power output of a certain system is 50 dBm. What is this power in watts?
4.3 Five networks in series have decibel gains as shown in Figure 4.9.
  Calculate the overall gain in dB and the power output at B if the input at A is 1 mW.
4.4 A 100 W amplifier has a linear power gain of 150. What must the input power be for the amplifier to give full output? What is the power gain of this amplifier in dB?

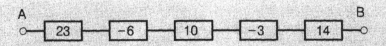

Figure 4.9 *Diagram for Question 4.3*

4.5 The frequency response of an amplifier is quoted as '100 Hz to 18 kHz (−3 dB)'.

A measurement of power output gives 20 W at 1 kHz, 15 W at 18 kHz and 14 W at 100 Hz. Is the amplifier within specification?

4.6 An attenuator introduces a loss of 12 dB. A voltage of 100 mV is applied to its input. What will be the voltage at the output?

5 Feedback

The concept of feedback

If part or all of the output signal is fed back and included with the input signal, the amplifier is said to possess feedback. When the signal is fed back deliberately by a known amount the results are predictable mathematically. If the signal fed back is by unintentional means or by an unknown amount, the results are less easy to calculate. The reader may have observed the effect of the output of an amplifier's loudspeaker being returned to the amplifier itself through a microphone. The result, a loud unpleasant squeal called 'howl round', is certainly predictable, but not worthy of calculation.

The principle of feedback can be considered by referring to the block diagram shown in Figure 5.1.

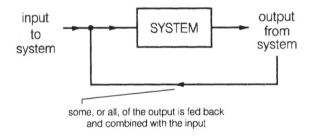

Figure 5.1 *The concept of feedback*

There are two types of feedback – negative and positive.

Important points

- **Negative feedback** is where the output signal fed back is **out of phase** with the input signal, resulting in a **reduction** in the overall input signal and in the gain of the amplifier.

- **Positive feedback** is where the output signal fed back is **in phase** with the input signal, resulting in an **increase** in the overall input signal and in the gain of the amplifier.

- Alternative names are degenerative feedback (negative) and regenerative feedback (positive), which are of opposite effect.

- The application of negative feedback is in applying a measure of **control** to an amplifier's performance.

- The application of positive feedback is to provide a prescribed **oscillation**, as for the signal generators used for the Practical Exercises.

Positive feedback and oscillators are dealt with in the following chapter. The remainder of this chapter is devoted entirely to negative feedback, with first a look at its effect on overall amplifier gain.

Open- and closed-loop gain

The block diagram of Figure 5.2(a) shows a system where the output signal voltage is totally dependent upon amplifier gain, with any changes in this gain resulting in corresponding changes in amplifier output.

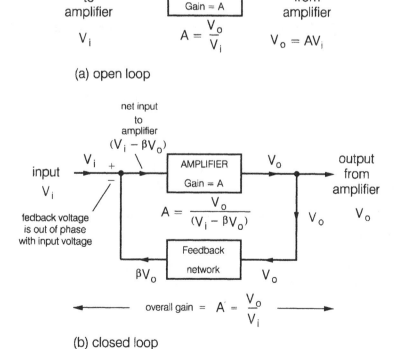

(a) open loop

(b) closed loop

Figure 5.2 *The open and closed loop amplifier*

The overall amplifier gain **without** feedback (A) is known as the **open-loop** gain. It can be seen that:

$$A = \frac{\text{output voltage of amplifier}}{\text{input voltage to amplifier}} = \frac{V_o}{V_i}$$

Figure 5.2(b) shows the amplifier with feedback. A fraction β (beta) of the output voltage is fed back out of phase, or in opposition to the input voltage. This opposition is indicated by the 'minus' sign. It therefore constitutes negative feedback and the overall gain, known as **closed-loop** gain, is now A'. It can be seen that the amplifier open-loop gain is still A, but the **net** input to the amplifier is equal to $(V_i - \beta V_o)$. The expression for A becomes:

$$A = \frac{V_o}{(V_i - \beta V_o)}$$

$A(V_i - \beta V_o) = V_o$ $\hspace{2cm}$ (by cross-multiplication)

$AV_i - A\beta V_o = V_o$ $\hspace{2cm}$ (multiplying out the bracket)

$AV_i = V_o + A\beta V_o$ $\hspace{2cm}$ (re-arranging)

$AV_i = V_o(1 + A\beta)$ $\hspace{2cm}$ (V_o is common to both terms)

$V_o(1 + A\beta) = AV_i$ $\hspace{2cm}$ (changing sides)

giving $\dfrac{V_o}{V_i} = \dfrac{A}{(1 + A\beta)}$ $\hspace{2cm}$ (dividing both sides by $(1 + A\beta)$)

We see from Figure 5.2(b) that the overall amplifier gain **with** feedback (A') is given by

$$A' = \frac{V_o}{V_i}$$

Thus we arrive at the expression for overall gain with feedback as

$$A' = \frac{V_o}{V_i} = \frac{A}{(1 + A\beta)}$$

To give this some meaning, let us work through an example to show at least one effect that negative feedback has on overall amplifier gain.

Important point

- The feedback fraction β is a **ratio** (for example $1/10$, $1/50$, 20% etc. of the output voltage) and is **not** a voltage.

Example 1

(a) Suppose that open-loop gain $A = 1000$
$\hspace{3cm}$ and feedback ratio $\beta = 1/10$

This will give an overall closed-loop gain A' of

$$\frac{1000}{(1 + 1000 \times 1/10)}, \text{ that is, } \frac{1000}{101}, \text{ or } 9.90$$

The first thing to notice is that the overall gain of the system has fallen dramatically, from 1000 to 9.90. Is there going to be any reward for this enormous sacrifice? We shall see.

(b) Suppose that A, for some reason, falls to 500, in other words by 50%.
β is still 1/10. Remember, it is a **ratio** of the output voltage.

The new value of overall gain with feedback is given by

$$A' = \frac{500}{(1 + 500 \times 1/10)}$$

$$= \frac{500}{51}$$

$= 9.80$, which amounts to a change of 1.0%

The second thing to notice is although the value of A has changed by 50%, the value of A' has changed by only 1.0%.

It is true of course that the overall gain with feedback (A') is very much reduced but in return for this, it will change very much less in percentage terms than the original change in amplifier gain (A).

Let us take this a stage further.

(c) The amplifier gain (A) now falls even further, to 250, which represents a total change of 75%.
(β is still 1/10)

The new value of overall gain with feedback is given by

$$A' = \frac{250}{(1 + 250 \times 1/10)}$$

$$= \frac{250}{26}$$

$= 9.61$, a total change of 2.9%

And so, even though the change in A is as much as 75%, the change in A' is still relatively small, at 2.9%.

A further point to be made, concerns the effect on the value of A' for larger values of A, with β remaining unchanged.

(d) The open-loop gain is 10 000 and β is 1/10. The table in Figure 5.3 shows the results of changes in A.

The conclusion here must be that the greater the sacrifice in gain, the greater the reward in terms of stability of gain.

From the expression

$$A' = \frac{A}{(1 + A\beta)}$$

A	% change in A	A'	% change in A'
10 000	——	9.99	——
5 000	−50	9.98	−0.1
2 500	−75	9.96	−0.3

Figure 5.3 *Negative feedback: stability of overall amplifier gain for Example 1(d)*

We can say that if $A\beta$ is very much larger than 1, then

$$(1 + A\beta) = A\beta \text{ (neglecting the '1')}$$

giving

$$A' = \frac{A}{A\beta}$$

$$= \frac{1}{\beta}$$

From the above example, with $\beta = 1/10$,

$$A' = \frac{1}{1/10} = 10$$

which is close to the value of 9.99 for A' obtained above!

Practical Exercise 5.1

To investigate the effect of negative feedback on gain and stability of a common emitter transistor amplifier.

For this exercise you will need the following components and equipment:

1 – npn transistor (BC109)
2 – resistor (470 Ω, 1 kΩ)
1 – resistor (10 kΩ, 47 kΩ)
3 – capacitor (100 µF, 16 V)
1 – DC power supply (+12 V)
1 – audio frequency signal generator
1 – double beam cathode ray oscilloscope

Procedure

1 Connect up the circuit shown in Figure 5.4(a). The pin connection diagram for the BC109 transistor is given in Figure 3.5.

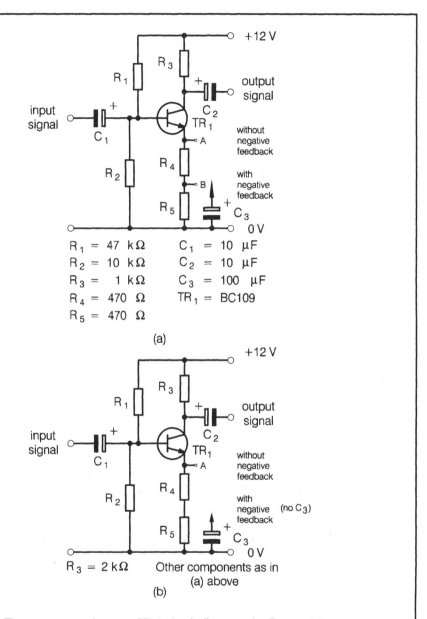

R₁ = 47 kΩ   C₁ = 10 μF
R₂ = 10 kΩ   C₂ = 10 μF
R₃ = 1 kΩ   C₃ = 100 μF
R₄ = 470 Ω   TR₁ = BC109
R₅ = 470 Ω

(a)

R₃ = 2 kΩ   Other components as in
(a) above
(b)

Figure 5.4   *The common emitter amplifier: circuit diagrams for Practical Exercise 5.1*

2 With capacitor C_3 in position A – 'without negative feedback' – and a signal input voltage of 10 mV peak-to-peak at 1 kHz, measure the voltage gain.
3 With the capacitor in position B – 'with negative feedback' – repeat the measurement of voltage gain, if necessary, with a larger input signal voltage.
4 For the circuit of Figure 5.4(b), repeat the measurement of voltage gain, without and with negative feedback.

Continued on p. 118

Practical Exercise 5.1 (*Continued*)

without feedback – C_3 to position *A*

with feedback – C_3 unconnected

5 Complete the table shown in Figure 5.5.

	Gain	
	Without negative feedback	With negative feedback
(a)		
(b)		

Figure 5.5   *Table of results for Practical Exercise 5.1*

6 Connect the CRO to the emitter of TR_1 and observe that the condition 'with negative feedback' causes a signal voltage to be present across the emitter resistors. This is the feedback voltage responsible for reducing the value of the overall gain.

Conclusions

1 Comparing the two circuits, calculate:
  (a) the percentage change in overall gain **without** negative feedback (this should be about −50%)
  (b) the percentage change in overall gain **with** negative feedback.

2 Give your views on the effectiveness of negative feedback in providing stability of overall gain.

The suggestion earlier was that the amplifier gain *A* could change 'for some reason'. What might this reason be?

In Practical Exercise 3.3, procedure 10, you investigated the effect on amplifier voltage gain of using transistors (BC107/108/109) with different values of current gain, h_{fe}. If all went according to plan your conclusions might have stated that the voltage gain depended very much on h_{fe}. As a result of this, in order to achieve a uniformity of amplifier performance for that particular circuit, it would be necessary to use transistors of the same type whose values of current gains were within a fairly close tolerance.

Important points

- The overall amplifier gain can be controlled by the use of negative feedback. The sacrifice is a much-reduced overall gain from the starting point with the reward being high **stability of gain**.

- The overall gain (A') becomes virtually **independent** of the amplifier gain (A) and dependent only on the value of the feedback ratio (β).

- The feedback ratio (β) is decided by resistor values and not by any transistor parameter.

Questions

5.1 An amplifier has a gain of 240. Calculate the overall gain when 1/120 of the output voltage is fed back in opposition to the input.

5.2 A certain amplifier has a gain of 180 without feedback and a gain of 18 with feedback. Calculate the feedback fraction.

5.3 A negative feedback amplifier has 0.5% of the output voltage fed back giving an overall gain of 160. Calculate the necessary value of the open-loop amplifier gain.

5.4 An amplifier with a voltage gain of 20 000 is used in a feedback circuit where $\beta = 0.02$. Calculate the overall gain.
 If the amplifier gain dropped by 50% of its initial value what would the overall gain now become?

5.5 In an amplifier having an input of 1 V, the output falls from 50 V to 25 V when feedback is applied. Calculate the fraction of the output that is fed back.
 Calculate the percentage reduction in overall gain, with and without feedback, if the amplifier gain fell from 50 to 40.

Practical Exercise 5.2

To investigate the effect of negative feedback on the frequency response of a common emitter transistor amplifier.
 For this exercise you will need the following components and equipment:

1 – npn transistor (BC109)
1 – resistor (470 Ω, 1 kΩ, 10 kΩ, 47 kΩ)
1 – capacitor (0.47 μF, 1000 pF)
1 – DC power supply (+12 V)
1 – audio frequency signal generator
1 – double beam cathode ray oscilloscope

 The full frequency response of the amplifier without the use of negative feedback has been covered in Chapters 3 and 4 and need not be repeated here. Furthermore

Continued on p. 120

Practical Exercise 5.2 *(Continued)*

the response with negative feedback will be similar and the comparison is shown in Figure 5.7.

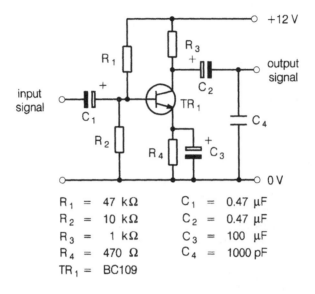

R_1 = 47 kΩ	C_1 = 0.47 μF
R_2 = 10 kΩ	C_2 = 0.47 μF
R_3 = 1 kΩ	C_3 = 100 μF
R_4 = 470 Ω	C_4 = 1000 pF
TR_1 = BC109	

Figure 5.6 *The common emitter amplifier: circuit diagram for Practical Exercise 5.2*

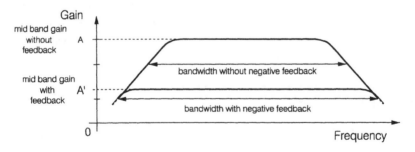

Figure 5.7 *Amplifier frequency response: the effect of negative feedback*

Important points

- This practical exercise will suggest a technique suitable for estimating the values of **mid band gain** and **bandwidth**.

- The addition of the 1000 nF capacitor will provide a convenient fall-off at high frequencies.

Procedure

1 Connect up the circuit shown in Figure 5.6, which is for the amplifier without negative feedback. The pin connection diagram for the BC109 transistor is given in Figure 3.5.
2 With a peak-to-peak input signal of 10 mV at 1 kHz, measure the voltage gain. This will be taken to be the mid band gain.
3 Vary the frequency below 1 kHz until the voltage gain has fallen to 0.707 of its mid band value. This will occur at frequency f_1.
4 Vary the frequency above 1 kHz until the voltage gain has fallen to 0.707 of its mid band value. This will occur at frequency f_2.
5 Calculate the bandwidth as $(f_2 - f_1)$.
6 Remove capacitor C_3 so that negative feedback is present.
7 Repeat procedures 2 to 5.
   The signal input voltage can be increased as necessary in order to obtain a worthwhile output level, but make sure the output waveform remains sinusoidal.

Conclusion

1 The actual mid band frequency response of an amplifier without negative feedback is rarely as flat and uniform as that suggested earlier. Explain how the use of feedback will produce a more flat and uniform response.

Important point

- The use of negative feedback produces a flatter, more uniform and wider frequency response at a lower overall gain.

Practical Exercise 5.3

To observe the effect of negative feedback on the distortion within a common emitter transistor amplifier.
   For this exercise you will need the following components and equipment:

1 – npn transistor (BC109)
2 – resistor (1 kΩ)
1 – resistor (470 Ω, 10 kΩ, 47 kΩ)
3 – capacitor (100 μF, 16 V)
1 – diode (1N4148)

Continued on p. 122

Practical Exercise 5.3 *(Continued)*

1 – potentiometer (1 kΩ)
1 – DC power supply (+12 V)
1 – audio frequency signal generator
1 – double beam cathode ray oscilloscope

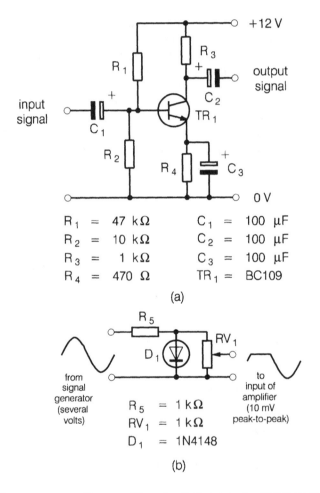

R_1 = 47 kΩ C_1 = 100 μF
R_2 = 10 kΩ C_2 = 100 μF
R_3 = 1 kΩ C_3 = 100 μF
R_4 = 470 Ω TR_1 = BC109

(a)

R_5 = 1 kΩ
RV_1 = 1 kΩ
D_1 = 1N4148

(b)

Figure 5.8 *The common emitter amplifier: circuit diagrams for Practical Exercise 5.3*

Procedure

1 Connect up the circuit shown in Figure 5.8(a). The pin connection diagram for the BC109 transistor is given in Figure 3.5.
2 With a signal input of 10 mV peak-to-peak at 1 kHz, measure the voltage gain.

3 Increase the input signal until the output signal **just** distorts (the positive peaks will start to round off). Sketch the output signal waveform and note its peak-to-peak value. At this point the voltage gain should be almost the same as that measured in 2 above.

Important point

- This distortion is being introduced **within the amplifier,** due to overloading of the amplifier and is called non-linear or **harmonic** distortion.

---

4 Introduce negative feedback by removing capacitor C_3. The voltage gain should fall as expected.
5 Increase the input signal until the peak-to-peak output signal voltage is the same as that measured in 3 above.
6 Sketch the output signal waveform and compare it with that drawn in 3 above. There should be a distinct difference.
7 To simulate distortion introduced with the input signal, modify the circuit to include that of Figure 5.8(b). Make sure that the capacitor C_3 is reconnected.
8 Note that the output signal is already distorted even at the low input signal level.
9 Introduce negative feedback, increase the input signal and note the continued distortion of the output signal.

Important points

- Negative feedback will assist in reducing non-linear distortion which is introduced **within the amplifier itself**.

- Negative feedback **will not** reduce distortion introduced from without the amplifier, that is, with an already distorted input signal.

- The effect on noise, which is 'unwanted signal' such as hiss and hum, will be exactly as for distortion, **providing** that the noise is introduced within the amplifier and not with the external signal.

- The reduction in distortion and noise is the same as that in the gain of the amplifier, which would suggest that the ratios of signal-to-distortion and signal-to-noise would be unchanged. However, since the distortion is most likely to occur in later, large-signal stages of amplification, negative feedback can be used over these stages, with the loss of gain being made up in earlier stages of the amplifier. To achieve best signal-to-noise ratio the earlier amplifier stages should use low-noise transistors or low-noise integrated circuits.

Feedback has to be **derived** at the output and it has to be **connected** from the output back to the input.

Types of derived feedback

(1) **Voltage** feedback is when the voltage fed back to the input is proportional to the output **voltage**.

(2) **Current** feedback is when the voltage fed back to the input is proportional to the output **current**.

The way the feedback is derived determines the resultant **output impedance** of the amplifier.

Methods of feedback connection

The voltage fed back may be connected to the input either in **series** or in **parallel** with the input.

The way the feedback is connected determines the resultant **input impedance** of the amplifier.

The diagram in Figure 5.9 illustrates the derivation and methods of connection.

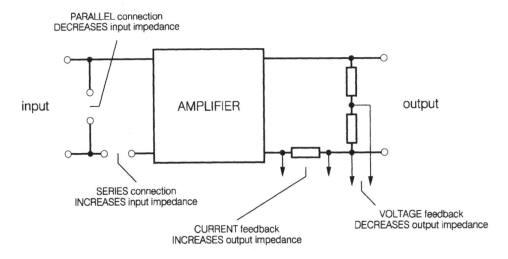

Figure 5.9 *Negative feedback: derivation and connection*

The derivation and the method of connection will be outlined in each of the practical circuits which follow.

Typical amplifier circuits employing negative feedback

(1)

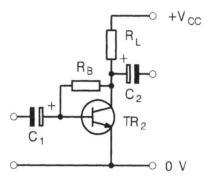

Figure 5.10 *Negative feedback: typical circuit 1*

For the circuit shown in Figure 5.10, the feedback voltage is developed across resistor R_B as a fraction of the output **voltage** between collector and 0 V, and is known as voltage feedback. It is connected across (in parallel with) the input, between base and 0 V and is a parallel connection. The description of the feedback method is usually in terms of 'input connection-output derivation', which for this circuit means **parallel-voltage**.

The feedback fraction, β, is given by $\beta = R_L/R_B$. So, for example, for $R_L = 4.7$ kΩ and $R_B = 100$ kΩ, $\beta \approx 1/20$ and the overall gain will be approximately 20.

Incidentally, the connection of R_B from collector to base also provides the DC bias current for the transistor, with the method being known as collector–base bias.

(2)

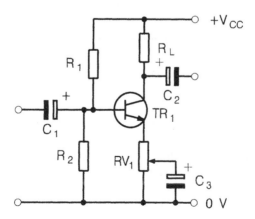

Figure 5.11 *Negative feedback: typical circuit 2*

The (well-known) circuit in Figure 5.11 shows a potentiometer (RV_1) as the emitter resistor with the bypass capacitor (C_3) being connected to the variable contact of the potentiometer. The part of RV_1 **not** in parallel with C_3 provides a feedback voltage across it proportional to the **current** through it. This is current feedback, which is then connected in **series** with the input. The feedback method is thus one of **series-current**.

The feedback fraction, β, is given by $\beta = R_L/R$ where R equals that part of RV_1 not in parallel with C_3. This circuit thus has its own built-in gain control which can at times be very useful.

To help the idea of the series connection, trace the path of the input circuit, from the upper input terminal on the left-hand side of C_1, to the base, emitter, through RV_1, along the 0 V line to the lower input terminal and finally through the input signal to the upper terminal where we began.

As to why the feedback voltage is negative, take the same route as above, this time considering an instantaneous polarity of the signal voltage at the input. Remembering that the signals at the input and emitter terminals will be **in phase** then you should agree that within the series input circuit, the input voltage and that across the emitter will be acting in opposite directions (anti-phase), that is, they will tend to cancel, producing a smaller net input to the amplifier.

The FET common source amplifier can be used in a similar fashion.

(3)

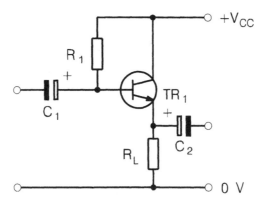

Figure 5.12   *Negative feedback: typical circuit 3*

For the circuit shown in Figure 5.12, known as an emitter follower, the output voltage is developed across the load resistor as usual, which this time is in the emitter. The feedback voltage developed across this resistor is equal to the output voltage since they are one and the same (and voltage derived) and connected in series with the input. The name **series-voltage** belongs here and the feedback is 100%.

The feedback fraction, β, is given by $\beta = R_L/R_L = 1$. This circuit will give an overall voltage gain of (almost) unity.

The field effect transistor source follower amplifier has a similar outcome.

Questions

5.6 Refer to Figure 5.13.

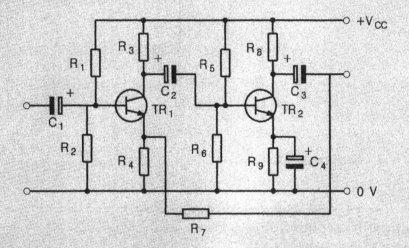

Figure 5.13 *Circuit diagram for Question 5.6*

   (a) Name the type of amplifier (common base, emitter or collector) used for each stage.

   (b) State the phase difference between the signal voltages at the collector of TR_2 and the base of TR_1.

   (c) Identify the feedback resistor(s) and hence explain how the feedback voltage is derived.

   (d) Explain how the feedback is connected to the input and hence give the 'input connection–output derivation' name.

   (e) State the expression for the feedback ratio β and the overall gain of the two-stage amplifier.

5.7 Refer to Figure 5.14.

   (a) Name the type of amplifier used for each stage.

   (b) Explain how the transistors TR_1 and TR_2 are biased.

   (c) State the phase difference between the signal voltages at the emitter of TR_2 and the base of TR_1.

   (d) Identify the feedback resistor(s) and hence explain how the feedback voltage is derived.

   (e) Explain how the feedback is connected to the input and hence give the 'input connection–output derivation' name.

   (f) State the expression for the feedback ratio β.

   (g) State the expression for the voltage gain of each stage and hence for the overall two-stage gain.

   (h) What is the name given to the coupling arrangement whereby there are no interstage coupling capacitors?

Continued on p. 128

Questions (*Continued*)

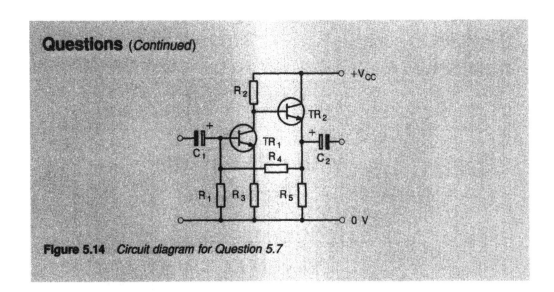

Figure 5.14   *Circuit diagram for Question 5.7*

Practical Exercise 5.4

To investigate the performance of a two-stage transistor amplifier with negative feedback.
For this exercise you will need the following components and equipment:

2 – npn transistor (BC109)
1 – resistor (470 Ω, 1 kΩ, 1.5 kΩ, 2.2 kΩ, 15 kΩ, 100 kΩ, 120 kΩ)
1 – capacitor (0.2 μF, 10 μF, 220 μF, all 35 V working)
1 – DC power supply (+20 V)
1 – audio frequency signal generator
1 – double beam cathode ray oscilloscope
1 – DC voltmeter

Important points

- This two-stage amplifier is directly coupled, since the collector of TR_1 goes **directly** to the base of TR_2, with no DC blocking capacitor.

- There are two feedback loops; one to provide negative feedback of the signal, and another to provide DC negative feedback to stabilize the DC operating conditions.

Procedure

1 Make up the circuit shown in Figure 5.15. The pin connection diagram for the BC109 transistor is given in Figure 3.5.

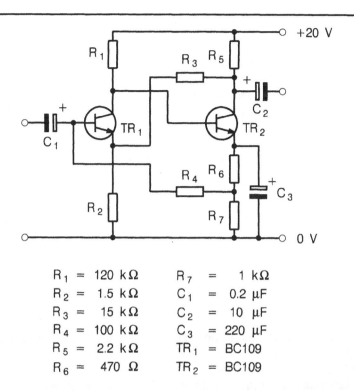

$$R_1 = 120 \text{ k}\Omega \qquad R_7 = 1 \text{ k}\Omega$$
$$R_2 = 1.5 \text{ k}\Omega \qquad C_1 = 0.2 \text{ μF}$$
$$R_3 = 15 \text{ k}\Omega \qquad C_2 = 10 \text{ μF}$$
$$R_4 = 100 \text{ k}\Omega \qquad C_3 = 220 \text{ μF}$$
$$R_5 = 2.2 \text{ k}\Omega \qquad TR_1 = \text{BC109}$$
$$R_6 = 470 \text{ Ω} \qquad TR_2 = \text{BC109}$$

Figure 5.15 *A two-stage amplifier: circuit diagram for Practical Exercise 5.4*

2 Measure the DC voltage at the base, emitter and collector of each transistor and confirm that both transistors are satisfactorily biased.
3 Apply the input signal and measure the overall voltage gain, remembering to maintain a non-distorted sinusoidal output signal.
  The value of this gain will be directly related to the feedback fraction β.
4 Use the CRO to establish the presence or otherwise of a signal voltage at the junction of R_6 and R_7 and at the emitter of TR_1. From this you should then be able to identify both of the feedback loops referred to above.

Conclusions

1 Which two resistors form the AC feedback loop?
2 Calculate the value of β and hence confirm your measurement of overall gain.
3 Which two resistors form the DC feedback loop?
4 This amplifier will have a higher input resistance than most of the amplifiers met earlier (approximately 100 kΩ) which enables a lower value (and non-electrolytic) capacitor to be used as C_1. Justify this reasoning by considering a potential divider network made up of a 0.2 μF capacitor in series with a 100 kΩ resistor with the output signal being across the resistor.

Important points

The advantages of using negative feedback in amplifiers can be summarized:

- The overall gain can be made virtually independent of stage gain and dependent **only** on the value of the feedback ratio β.

- A more uniform frequency response with a wider bandwidth.

- Noise and distortion introduced **within** the amplifier is reduced.

- The values of the input and output impedance can be controlled by the choice of feedback network.

- Additional phase shift, introduced by external coupling capacitors and resistors, is reduced.

Questions

5.8 In a certain amplifier, all of the output voltage is fed back in anti-phase with the input voltage. What will the overall voltage gain be and what is the name given to this type of amplifier?

5.9 Four amplifiers are made using the BC109 transistor, with each transistor having a different value of h_{fe}, as follows: 200, 400, 600 and 800.
   The collector load and input resistance of each amplifier have a value of 4.7 kΩ and 1 kΩ respectively.
   Calculate the open-loop voltage gain of each amplifier.

   Negative feedback is now employed having a feedback fraction of 1/20.
   Calculate the closed-loop voltage gain of each amplifier and the percentage increase in voltage gain (from minimum to maximum) with and without feedback.

5.10 An amplifier has a gain of 60 dB without feedback and 20 dB after feedback is applied. The gain without feedback now changes to 70 dB. What will be the new gain with feedback?
   Hint. The (linear) negative feedback formula **cannot** be used with dB units. Convert the dB values into linear units.

5.11 The open-loop gain of a range of manufactured amplifiers is found to vary from 700 to 400. Compare this percentage variation with that obtained when negative feedback is applied using a feedback factor of (a) 0.01, (b) 0.1.

5.12 'The use of negative feedback for stabilization of gain is a "trade-off" situation.' Explain this statement.

6 | Operational amplifiers

As the name suggests, the operational amplifier, or 'op-amp', is capable of performing mathematical operations, which are defined later in this chapter. It was developed originally for use in analog computers and used discrete components, in particular thermionic valves. The associated problems of heat dissipation and bulk restricted its general use. The advent of the transistor (1948+) and the integrated circuit (1960+) provided an entry for the operational amplifier into virtually all areas of electronics.

Definition of an operational amplifier

The operational amplifier is a high-gain, directly coupled voltage amplifier, having a frequency response which extends to DC (0 Hz) by virtue of the absence of coupling capacitors. The open-loop voltage gain is typically of the order of 100 000 (100 dB) and the op-amp is almost always used with a feedback network to provide a variety of functions.

The diagram in Figure 6.1 shows the symbol for the operational amplifier.

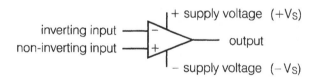

Figure 6.1 *The operational amplifier: symbol*

Important points

- There are **two** inputs:
  (i) one **inverting** input (shown as '−') where the output signal will be the opposite polarity or phase to the input signal
  (ii) one **non-inverting** input ('+') where the output signal will be the same polarity or phase as the input signal.

- There is **one** output.

Continued on p. 132

Important points (*Continued*)

- The basic operation is always as a **difference**, or differential amplifier, in which the output voltage is proportional to the difference between the two input voltages.

- A dual polarity power supply ($\pm V_S$) is normally used in order that the output voltage can vary either side of zero.

Characteristics of the operational amplifier

The **ideal** op-amp would have:

(a) infinite open-loop voltage gain
(b) infinite input impedance (giving no loading effect on signal source)
(c) zero output impedance (no internal losses)
(d) infinite bandwidth (all frequencies amplified equally)
(e) zero output for equal inputs.

A tall order! Practical values, of course, fall short of these demands, but not by a troublesome amount, particularly where both bipolar and field-effect integrated circuits are available. Let us take a look at these characteristics in more detail. The typical values given relate to the bipolar 741 ic.

Open-loop voltage gain

> This is always specified as low frequency gain since the 'gain–bandwidth product' must be taken into account
> Typical value: 200 000 times, or 106 dB

Closed-loop voltage gain

> The use of feedback (negative for amplifiers) will provide predictable values to suit the particular requirement.

Input impedance

> The impedance (often just resistance) between the two input terminals
> Typical value: 2 MΩ

Output impedance

> The output impedance (resistance) on open loop
> Typical value: 75 Ω

Gain–bandwidth product and transition frequency

> The gain–bandwidth product for any amplifier is the linear voltage gain multiplied by the bandwidth at that gain.

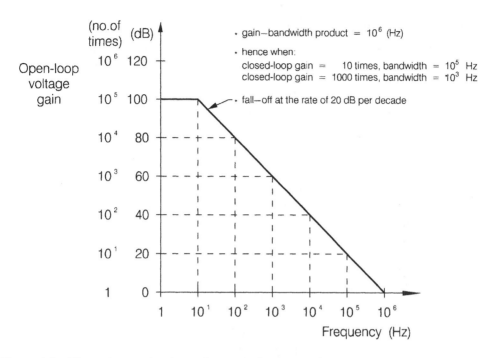

Figure 6.2 *The 741 operational amplifier: typical open-loop/gain frequency characteristic*

The typical relationship between **open-loop** voltage gain and bandwidth for an operational amplifier is shown in Figure. 6.2, where the low frequency gain is shown as 100 dB.

The open-loop gain falls with increasing frequency at a rate of 6 dB per octave (a doubling of frequency) equivalent to a rate of 20 dB per decade (10 × frequency).

The value of frequency at which the open-loop gain is equal to unity is called the transition frequency (f_T). For the example shown, f_T equals 1 MHz.

The bandwidth of the amplifier on **closed-loop** thus depends upon the values of gain and f_T.

For the example shown, a gain of 20 dB (10 times) will give a bandwidth of 100 kHz, whereas a gain of 60 dB (1000 times) will restrict the bandwidth to 1 kHz. And so,

$$\text{bandwidth} = \frac{f_T}{\text{closed-loop voltage gain (number of times)}}$$

Typical value: 1 MHz

Input offset voltage

For the ideal operational amplifier, with both inputs at zero, there should be zero output. This may not always be the case, due to imbalances within the amplifier, thereby resulting in a small output voltage. This effect can be nullified by applying a small offset voltage to the amplifier. Happily this is not normally a problem if external capacitive coupling is used.

Common mode rejection ratio (CMRR)

The ideal operational amplifier should give zero output signal when both inputs are supplied with identical signals (this is described as being in **common mode**).

$$CMRR = \frac{\text{gain with differential signals}}{\text{gain with common mode signals}}$$

Typical value: 90 dB

Slew rate

This is the rate of change of output voltage following a step (rectangular) input voltage. It is of particular importance when large pulse amplitudes are present. The situation is illustrated by the diagram in Figure. 6.3.

Typical value: 0.5 V/μs

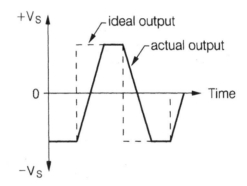

Figure 6.3 *The effect of slew rate*

Supply voltage range

Dual polarity supply, with minimum and maximum limits for satisfactory operation.

Typical values: ±5 V to ±18 V

The inverting amplifier

A reminder that the variable DC input signal voltage can be easily obtained from the main ±15 V supply, as shown in Figure 1.4.

Practical Exercise 6.1

To investigate the action of an inverting operational amplifier.

For this exercise you will need the following components and equipment:

1 – ic operational amplifier (741)
2 – resistor (1 kΩ)
1 – resistor (2.2 kΩ, 4.7 kΩ, 10 kΩ)
1 – variable resistor (4.7 kΩ)
2 – capacitor (10 μF 16 V)
1 – DC power supply (±15 V)
1 – DC voltmeter
1 – audio frequency signal generator
1 – double beam cathode ray oscilloscope

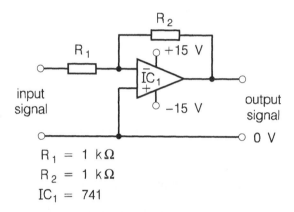

$R_1 = 1\ k\Omega$
$R_2 = 1\ k\Omega$
$IC_1 = 741$

Figure 6.4 *The inverting operational amplifier: circuit diagram for Practical Exercise 6.1 (Procedures (a) and (b))*

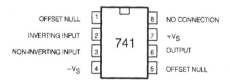

Figure 6.5 *The 741 operational amplifier: pin connections*

Procedure (a) Measurement of closed-loop gain using a DC signal

1 Connect up the circuit shown in Figure 6.4. The pin connections for the 741 ic are given in Figure 6.5.
2 Refer to the table in Figure 6.6. With the appropriate values of R_2 and V_i, use the voltmeter to measure the output voltage, taking care to observe the polarity of this voltage. Complete the table.

Continued on p. 136

Practical Exercise 6.1 (*Continued*)

		Measure		Calculate		Measure
R_1 (kΩ)	R_2 (kΩ)	V_i (V)	V_o (V)	$\dfrac{R_2}{R_1}$	$\dfrac{V_o}{V_i}$	Voltage * at (−) input
1	1	+1				
1	2.2	+1				
1	4.7	+1				
1	10	+1				
1	1	−1				
1	2.2	−1				
1	4.7	−1				
1	10	−1				

* expect this
to be very
small

Figure 6.6 *The inverting operational amplifier: table of results for Practical Exercise 6.1 (Procedure (a))*

Procedure (b) Measurement of transfer characteristic

1 With $R_2 = 4.7$ kΩ, vary V_i as shown in the table given in Figure 6.7, and measure the corresponding value of V_o. Complete the table.

V_i (V)	+5	+4	+3	+2	+1	0	−1	−2	−3	−4	−5
V_o (V)											

Figure 6.7 *The inverting operational amplifier: table of results for Practical Exercise 6.1 (Procedure (b))*

2 From the results of Procedure (b)1, draw the output–input transfer characteristic, which should resemble that shown in Figure 6.8.

Conclusions

1 From Procedure (a)2, compare the calculated values of V_o/V_i with those for R_2/R_1. Are they equal (or very nearly)?
2 Is the polarity of the output voltage always opposite to that of the input? Is this therefore an inverting or non-inverting amplifier?

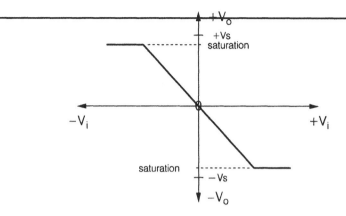

Figure 6.8 *The inverting operational amplifier: the output–input (transfer) characteristic*

3 From the characteristic drawn in Procedure (b)2, notice that the output voltage reaches a limit once a certain input voltage has been reached.
   Estimate the value of this limit, and the value of the input voltage at which it first occurs.
4 Comment on the measured values of the voltage at the inverting input terminal (the junction of R_1 and R_2).

Important points

For the inverting amplifier:

• The closed-loop voltage gain (A') is given by

$$A' = -\frac{R_2}{R_1}$$

where the 'minus' sign shows an **inversion**.

• The amplifier output reaches a limit (called **saturation** value) at approximately 2 V below the value of the supply, that is, at ±13 V.

• The resistor R_2 provides **negative** feedback by feeding a fraction of the out-of-phase output voltage back to the input.

Procedure (c) Measurement of closed-loop gain using an AC signal

1 Connect up the circuit of Figure 6.9.
2 Set the signal generator output to 1 V peak-to-peak at 1 kHz.

Continued on p. 138

Practical Exercise 6.1 *(Continued)*

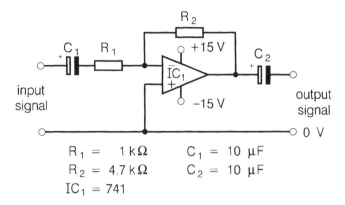

$$R_1 = 1\,k\Omega \qquad C_1 = 10\ \mu F$$
$$R_2 = 4.7\,k\Omega \qquad C_2 = 10\ \mu F$$
$$IC_1 = 741$$

Figure 6.9   *The inverting operational amplifier: circuit diagram for Practical Exercise 6.1 (Procedure (c))*

3 Display both input and output waveforms on the cathode ray oscilloscope and check that the output voltage is also sinusoidal and an inversion of the input. Measure the voltage gain of the amplifier and confirm agreement with the theoretical value.
4 Check the voltage level of the signal present at the inverting input terminal (the junction of R_1 and R_2).
5 Steadily increase the input voltage and notice that eventually the output voltage will 'square off', which is the saturation level referred to earlier.
   Check that this ± level has the same value as found in Procedure (b)2.
6 Connect the 4.7 kΩ variable resistor in series with the feedback resistor R_2. This will now enable the voltage gain to be varied between certain limits, with the variable resistor acting as a 'volume control'.
7 Using a signal input voltage of 1 V peak-to-peak, measure this available range of voltage gain.

Conclusion

1 Comment on your result for Procedure (c)4.

Important point

• The circuit of an inverting amplifier will, in practice, include a resistor between the non-inverting input and the 0 V line, as shown in Figure 6.10 by resistor R_3. Its purpose is to help to minimize any possible offset voltage, with its value being equal to the parallel combination of R_2 and R_1.

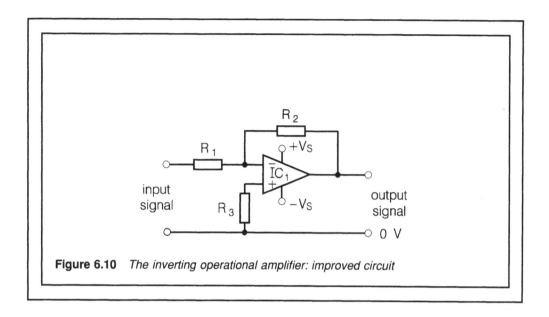

Figure 6.10   *The inverting operational amplifier: improved circuit*

Further analysis of the inverting amplifier

A block diagram of the inverting amplifier is given in Figure 6.11 and shows the signal voltages and currents under normal working conditions. To explain further:

(a) the current i_1 through the series component R_1 is a result of the input voltage v_i
(b) the feedback current i_f through the feedback component R_f is a result of the output voltage v_o
(c) the currents i_1 and i_f meet at the junction of R_1 and R_f, which is the inverting input terminal
(d) the voltage at the inverting input terminal is v_{in}.

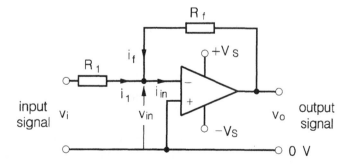

Figure 6.11   *The inverting operational amplifier: circuit diagram for further analysis*

It would be helpful to remember that the operational amplifier has:

(a) high gain (typically $A = 10^5$)
(b) high input resistance (typically $R_{in} = 2$ MΩ).

The first thing to do in this analysis is to consider the situation at the junction of R_1 and R_f. On the basis that the current entering the junction is equal to the current leaving – Kirchhoff's second Law – we can state that:

$$i_1 + i_f = i_{in}$$

(This is called 'summing the currents at the junction')

where $\quad i_1 = \dfrac{v_i - v_{in}}{R_1} \quad$ and $\quad i_f = \dfrac{v_o - v_{in}}{R_f}$

thus $\quad \dfrac{v_i - v_{in}}{R_1} + \dfrac{v_o - v_{in}}{R_f} = i_{in}$

Since the amplifier input resistance (R_{in}) is very high, the input current will be very small.

The **first assumption** we make is that this current is small enough to be neglected, that is, it can be regarded as being zero. Consequently, the above expression becomes:

$$\dfrac{v_i - v_{in}}{R_1} + \dfrac{v_o - v_{in}}{R_f} = 0$$

We know that $v_o = -Av_{in}$, which after re-arranging, gives:

$$v_{in} = -\dfrac{v_o}{A}$$

Now, substituting for v_{in} and taking care with the 'minus' signs, we get:

$$\dfrac{v_i + \frac{v_o}{A}}{R_1} + \dfrac{v_o + \frac{v_o}{A}}{R_f} = 0$$

Since the amplifier gain (A) is very high, then the value of v_o/A will be very much smaller than both v_i and v_o.

Thus the **second assumption** we make is that v_o/A can be neglected compared with v_i and v_o. The next stage is to say that

$$\dfrac{v_i}{R_1} + \dfrac{v_o}{R_f} = 0$$

and $$\dfrac{v_o}{R_f} = -\dfrac{v_i}{R_i}$$

which becomes $$\dfrac{v_o}{v_i} = -\dfrac{R_f}{R_i}$$

A very important aspect of the inverting operational amplifier concerns the actual voltage at the inverting input terminal. The assumption was made (and hopefully accepted) that the voltage given by v_o/A was small enough to be neglected.

This voltage was the input voltage (v_{in}) at the inverting input terminal. The measurement in Practical Exercise 6.1, Procedures (b) and (c) should now be confirmation that the assumption was valid. The inverting input is almost, or **virtually,** at zero potential and is said to be a **virtual earth**. It is of course not actually connected to earth and further thinking will provide the realization that the input voltage cannot be absolutely zero, since if it were, then the output would be zero also. Let us content ourselves with the fact that this voltage **is** very small and the internal gain very high, so that a prescribed output will be achieved for a particular situation.

Important points

- The overall gain of the inverting amplifier is dependent only upon resistor values, and independent of the internal amplifier gain.
  (It would be helpful to recall that this is another example of negative feedback at its best!)

- The 'minus' sign attached to the right-hand side shows the polarity or phase inversion.

- The inverting input is a virtual earth.

Questions

Questions 6.1 to 6.3 refer to the amplifier shown in Figure 6.4.

6.1 If $R_1 = 1$ kΩ, and $R_2 = 2$ kΩ, calculate the output voltage when
  (a) the input voltage is $+0.5$ V
  (b) the input voltage is -1.0 V

6.2 $R_1 = 1$ kΩ. Calculate the necessary value of R_2 in order for the output to be ten times the input voltage.

6.3 The gain is required to be fully variable between a minimum of -1 and a maximum of -11. Suggest how this might be achieved, sketching a suitable circuit and calculating component values.
  What restrictions are there for the input signal voltage when the amplifier is at maximum gain?

The non-inverting amplifier

The circuit diagram of a non-inverting amplifier is shown in Figure 6.12.

The input is applied to the non-inverting input terminal with the negative feedback from the output being connected to the inverting input terminal. The differential input principle holds good and as before, the voltage difference between the two inputs (v_{in}) is very small.

We can state that

$$v_i - \frac{R_1}{R_1 + R_f} v_o = v_{in} = 0$$

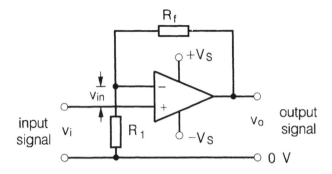

Figure 6.12 *The non-inverting operational amplifier: circuit diagram for further analysis*

thus

$$v_i = \frac{R_1}{R_1 + R_f} v_o$$

giving

$$\frac{v_o}{v_i} = \frac{R_1 + R_f}{R_1} = 1 + \frac{R_f}{R_1}$$

Practical Exercise 6.2

To investigate the action of a non-inverting AC operational amplifier.
  For this exercise you will need the following components and equipment:

1 – ic operational amplifier (741)
2 – resistor (1 kΩ)
1 – resistor (2.2 kΩ, 4.7 kΩ, 10 kΩ, 100 kΩ)
2 – capacitor (10 μF, 16 V)
1 – DC power supply (±15 V)
1 – audio frequency signal generator
1 – double beam cathode ray oscilloscope

Procedure (a) Voltage gain

1 Connect up the circuit shown in Figure 6.13(a). The pin connections for the 741 ic are shown in Figure 6.5.
2 Set the signal generator output to 1 V peak-to-peak at 1 kHz and measure the amplifier output voltage. Calculate the voltage gain.
3 Repeat the measurement of voltage gain for $R_2 = 2.2$ kΩ, 4.7 kΩ, and 10 kΩ.
4 Complete the table shown in Figure 6.14.
5 Note the phase relationship between output and input voltages.
6 Check that saturation of the output signal occurs as found previously for the inverting amplifier.

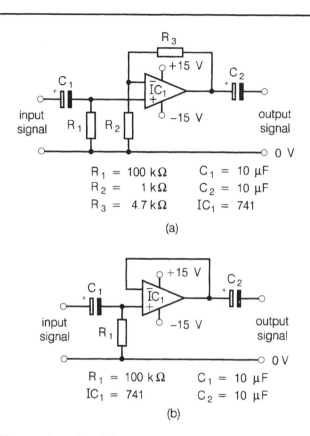

$$R_1 = 100\ \text{k}\Omega \qquad C_1 = 10\ \mu\text{F}$$
$$R_2 = \quad\ 1\ \text{k}\Omega \qquad C_2 = 10\ \mu\text{F}$$
$$R_3 = \ 4.7\ \text{k}\Omega \qquad IC_1 = 741$$

(a)

$$R_1 = 100\ \text{k}\Omega \qquad C_1 = 10\ \mu\text{F}$$
$$IC_1 = 741 \qquad C_2 = 10\ \mu\text{F}$$

(b)

Figure 6.13 *The non-inverting AC operational amplifier: circuit diagrams for Practical Exercise 6.2*

		Measure		Calculate	
R_2 (kΩ)	R_3 (kΩ)	v_i (V)	v_o (V)	$\dfrac{R_2 + R_3}{R_3}$	$\dfrac{V_o}{V_i}$
1	1	1			
1	2.2	1			
1	4.7	1			
1	10	1			
1	1	1			
1	2.2	1			
1	4.7	1			
1	10	1			

Figure 6.14 *The non-inverting AC operational amplifier: table of results for Practical Exercise 6.2 (Procedure (a))*

Continued on p. 144

Practical Exercise 6.2 (*Continued*)

Procedure (b) The voltage follower

1 Connect up the circuit shown in Figure 6.13(b).
2 With a frequency of 1 kHz, measure the voltage gain and the phase difference between output and input voltages over a range of input voltages.

Conclusion

1 Comment on the voltage gain and the phase shift of the amplifier used in Procedure (b). Compare their similarity with the results from the emitter follower amplifier in Practical Exercise 3.6.
  Hence suggest a practical application for this type of non-inverting amplifier.

Important point

- For the non-inverting **AC amplifiers** of Figure 6.13, the resistor R_1 acts as a DC return for the non-inverting input. For non-inverting **DC amplifiers**, where coupling capacitors are absent, R_1 will be unnecessary.

Questions

Questions 6.4 to 6.6 refer to the amplifier shown in Figure 6.15.

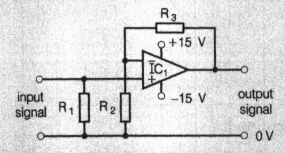

Figure 6.15 *Circuit diagram for Questions 6.4 to 6.6*

6.4 Calculate the voltage gain when $R_2 = 4.7$ kΩ and $R_3 = 10$ kΩ and the output voltage for an input of −0.2 V.
6.5 An output of ±1.44 V is required from an input of ±30 mV. Calculate suitable values for R_2 and R_3.

6.6 A non-inverting amplifier is to provide a voltage gain, variable from 1 to 10, by replacing R_3 with a variable resistor. Suggest a suitable (practical) value for this resistor given that the value of R_2 is 4.7 kΩ.

Use your practical value to estimate the actual range of voltage gain.

The summing amplifier

The summing amplifier is an extension of the inverting amplifier met earlier and from the circuit in Figure 6.16 we can see that each input has a prescribed gain as follows:

$$\text{gain of input } 1 = -\frac{R_4}{R_1}$$

$$\text{gain of input } 2 = -\frac{R_4}{R_2}$$

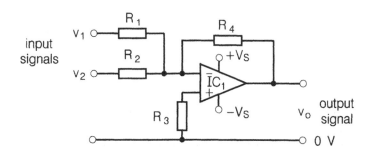

Figure 6.16 *The summing operational amplifier: the basic circuit*

The output voltage v_0 is made up of contributions from each input and as a result, the general expression for v_0 will be given by

$$v_0 = -\left(\frac{R_f}{R_1}v_1 + \frac{R_f}{R_2}v_2 + \text{ etc.}\right)$$

(Do not forget the 'inversion'.)

Practical Exercise 6.3

To investigate the action of a summing operational amplifier.
For this exercise you will need the following components and equipment:

1 – ic operational amplifier (741)
2 – resistor (1 kΩ)
1 – resistor (2.2 kΩ, 4.7 kΩ)

Continued on p. 146

Practical Exercise 6.3 *(Continued)*

1 – DC power supply (±15 V)
1 – DC voltmeter

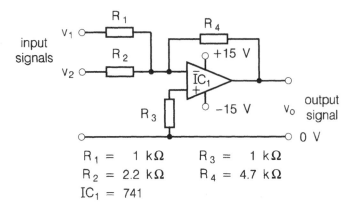

$$R_1 = 1\ k\Omega \qquad R_3 = 1\ k\Omega$$
$$R_2 = 2.2\ k\Omega \qquad R_4 = 4.7\ k\Omega$$
$$IC_1 = 741$$

Figure 6.17   *The summing operational amplifier: circuit diagram for Practical Exercise 6.3*

Procedure

1 Connect up the circuit of Figure 6.17. The pin connections for the 741 ic are shown in Figure 6.5.
2 With an input voltage of +1 V to one input only, measure the output voltage, and hence the gain, **of each input separately**.
3 With a voltage of +1 V to each input, measure the total output voltage.
4 Repeat the above measurements using input voltages of your choice, both positive and negative, to confirm the action of the summing amplifier.

Questions

Questions 6.7 to 6.9 refer to the amplifier shown in Figure 6.16.

6.7 $R_1 = 1\ k\Omega$, $R_2 = 1\ k\Omega$, $R_4 = 2.2\ k\Omega$.
  $v_1 = +0.1$ V, $v_2 = +0.2$ V.
  Calculate the value of the output voltage v_0.
6.8 $R_1 = 1\ k\Omega$, $R_2 = 2.2\ k\Omega$, $R_4 = 4.7\ k\Omega$.
  $v_1 = -0.5$ V, $v_2 = +0.5$ V.
  Calculate the value of the output voltage v_0.
6.9 Figure 6.18 shows the waveforms of the input voltages v_1 and v_2 respectively. If each input has unity gain, sketch the waveform of the output voltage.

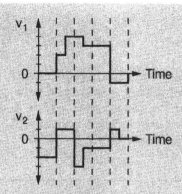

Figure 6.18 *Waveform diagrams for Question 6.9*

Hint. The waveforms should be added together point by point taking into account their respective polarities.

6.10 A summing amplifier is required to give an output voltage v_0, where v_0 is given by

$$v_0 = -(0.5v_1 + 2.5v_2 + 5.0v_3)$$

(a) Starting from a choice of feedback resistor, calculate values for the input resistors.
(b) Calculate the output voltage, when

$$v_1 = +1 \text{ V}, \quad v_2 = +0.4 \text{ V and } v_3 = -0.4 \text{ V}$$

The difference amplifier

As mentioned earlier, all operational amplifiers work on the difference principle but the circuits previously discussed have been 'single-ended', that is, the input signal has been applied to one or other of the input terminals but not to both at the same time. The circuit of the differential amplifier is shown in Figure 6.19.

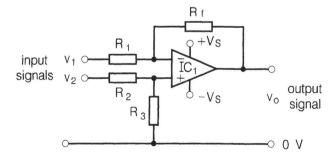

Figure 6.19 *The difference operational amplifier: the basic circuit*

The output voltage (v_o) is given by

$$v_o = -\frac{R_f}{R_1}(v_1 - v_2) \qquad \text{(remember the 'inversion')}$$

This type of amplifier finds application in instrumentation systems, being particularly useful in bridge circuits. Its advantage is that equal in-phase signals at the two inputs produce no output. The input resistance values will be R_1 for the inverting input and $(R_2 + R_3)$ for the non-inverting input. Since, normally, $R_1 = R_2$, these values will be unequal and therein lies a disadvantage of this circuit.

Practical Exercise 6.4

To investigate the action of a differential operational amplifier.
   For this exercise you will need the following components and equipment:

1 – ic operational amplifier (741)
4 – resistor (10 kΩ)
1 – DC power supply (±15 V)
1 – DC voltmeter

Procedure

1 Connect up the circuit of Figure 6.20. The pin connections for the 741 ic are shown in Figure 6.5.

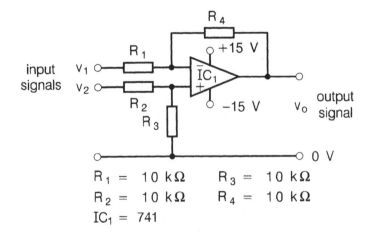

$$R_1 = 10 \text{ k}\Omega \qquad R_3 = 10 \text{ k}\Omega$$
$$R_2 = 10 \text{ k}\Omega \qquad R_4 = 10 \text{ k}\Omega$$
$$IC_1 = 741$$

Figure 6.20   *The differential operational amplifier: circuit diagram for Practical Exercise 6.4*

2 With input voltages v_1 and v_2 both positive, measure the output voltage v_0 and note its polarity, with

(a) v_1 more positive than v_2
(b) v_2 more positive than v_1.

3 With input voltages v_1 and v_2 both negative, measure the output voltage v_0 and note its polarity, with

(a) v_1 more negative than v_2
(b) v_2 more negative than v_1.

4 Try to make voltages v_1 and v_2 **exactly equal** and of the same polarity. Measure the output voltage.

Conclusions

1 Complete the following statements by choosing the correct word in each bracket:

(a) When input v_1 is more positive than input v_2, the output v_0 is (*positive/negative*).
(b) When input v_1 is more negative than input v_2, the output v_0 is (*positive/negative*).
(c) With equal input voltages of the same polarity, the output voltage should be equal to (*+15 V/–15 V/0 V*).

2 What is the voltage gain of each input?
3 Why is it necessary for resistors R_1 and R_2 to be equal?

Question

Question 6.11 refers to the amplifier shown in Figure 6.20.

6.11 $R_1 = R_2 = 10$ kΩ and $R_4 = 47$ kΩ. The voltages applied to the two inputs are 100 mV and 150 mV respectively.
   Calculate the output voltage.

The comparator

This is a device (see Figure 6.21) in which the operational amplifier is used on open-loop, that is, without negative feedback. Since the open-loop gain is very high (typically 100 000), a very small voltage difference between the two inputs will be sufficient to cause the output voltage to go into saturation. It will then be in one or other of two states; either just below the positive supply or just below the negative supply.

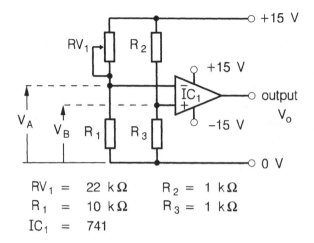

$$RV_1 = 22 \text{ k}\Omega \qquad R_2 = 1 \text{ k}\Omega$$
$$R_1 = 10 \text{ k}\Omega \qquad R_3 = 1 \text{ k}\Omega$$
$$IC_1 = 741$$

Figure 6.21 *The comparator operational amplifier: circuit diagram for Practical Exercise 6.5*

Practical Exercise 6.5

To investigate the action of a comparator operational amplifier.
    For this exercise you will need the following components and equipment:

1 – ic operational amplifier (741)
2 – resistor (1 kΩ)
1 – resistor (10 kΩ)
1 – variable resistor (22 kΩ)
1 – light-dependent resistor (ORP 12)
1 – DC power supply (±15 V)
1 – DC voltmeter

Procedure

1 Connect up the circuit shown in Figure 6.21. The pin connections for the 741 ic are given in Figure 6.5.
2 Measure the voltage V_B at the non-inverting input. For equal values of R_2 and R_3 it should be one-half of the supply voltage.
3 Adjust RV_1 so that the voltage V_A is larger than the voltage V_B. Note the value and polarity of the output voltage V_o.
4 Adjust RV_1 so that the voltage V_B is larger than the voltage V_A. Note the value and polarity of the output voltage V_o.
    What will the output voltage be when V_A and V_B are equal?
5 Try it and see. The difficulty will be in getting these two voltages to be **exactly equal**, with any slight inequality resulting in the output switching from one saturated state to the other, very quickly!

6 Replace RV_1 with the light-dependent resistor (LDR).

Important points

- When the inverting input is more positive than the non-inverting input, the output will be negative.

- When the non-inverting input is more positive than the inverting input, the output will be positive.

- The light-dependent resistor is a photo conductive cell having a high resistance when covered ('dark' resistance $= 1$ MΩ) and a low resistance when exposed to light (typically 6 kΩ).

---

7 With the LDR exposed to light, check that the output voltage is negative.
8 Cover the LDR and watch the output voltage switch positive.

Conclusion

1 Suggest a practical application for the LDR circuit above. What additional components might be necessary in order for this application to be effective?

Frequency limitations of the operational amplifier

An important characteristic of an operational amplifier is the quantity known as slew rate. Slew rate is important at high frequency operation since high frequency means rapid change. As shown in Figure 6.22(a) the effect of slewing is to cause the output voltage to change at a slower rate than the input. The resulting output waveform is a **distortion** of the input.

Important points

- Slew rate is measured in terms of the time (t) taken for the output voltage to go from 10% to 90% of its final value (V_{max}), following a sudden change or **step** in the input voltage. See Figure 6.22(b).

- Slew rate (S) is given by

$$S = \frac{0.9V_{max} - 0.1V_{max}}{t} \frac{\text{(V)}}{\text{(s)}}$$

Continued on p. 152

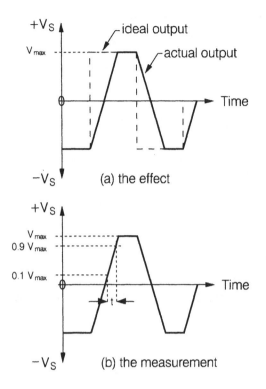

(a) the effect

(b) the measurement

Figure 6.22 *Slew rate*

Important points (*Continued*)

- The time 't' will typically be measured in microseconds and so the units for slew rate will be volts per microsecond (V/µs).

- From the measurement of slew rate, we can determine the maximum frequency of operation for sinusoidal signals, f_{max}, otherwise known as the **power bandwidth**, as:

$$f_{max} = \frac{S}{2\pi V_p}$$

where V_p is the peak output voltage in volts.

- The frequency f_{max} can be increased by
  (a) reducing the output voltage
  (b) using an ic having a higher slew rate.

- The power bandwidth will be less than the small-signal bandwidth calculated from the gain–bandwidth product.

Example 1

The slew rate of a certain bipolar ic operational amplifier is 0.5 V/μs. Calculate the maximum undistorted sinusoidal frequency for a **peak** voltage output of:

(a) 10 V
(b) 5 V.

(a) $V_p = 10$ V

$$f_{max} = \frac{S}{2\pi V_p}$$

$$= \frac{0.5 \times 10^6}{2\pi \times 10} \quad \text{(converting } S \text{ into V/s)}$$

$$= 7957 \text{ Hz}$$

(b) $V_p = 5$ V
Halving the peak output voltage V_p will double the frequency f_{max}, thus

$$f_{max} = 15915 \text{ Hz} = 15.9 \text{ kHz}$$

This higher value of frequency could be achieved with a 10 V peak output voltage with an ic having a slew rate of 1 V/μs.

Practical Exercise 6.6

To investigate the slew rate of an operational amplifier.
For this exercise you will need the following components and equipment:

1 – ic operational amplifier (741, LF351)
1 – resistor (100 kΩ)
2 – capacitor (10 μF, 16 V)
1 – DC power supply (±15 V)
1 – audio frequency signal generator (sine and square)
1 – double beam cathode ray oscilloscope

Procedure

1 Connect up the circuit shown in Figure 6.23. The pin connection diagram for the 741 ic is shown in Figure 6.5.
2 Set the output voltage of the generator to give a 10 V **square** wave at 10 kHz. Observe the input and output voltage waveforms with the CRO. They should be similar to those shown in Figure 6.22.

Continued on p. 154

Practical Exercise 6.6 (*Continued*)

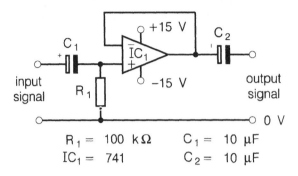

$R_1 = 100 \ k\Omega$ $C_1 = 10 \ \mu F$
$IC_1 = 741$ $C_2 = 10 \ \mu F$

Figure 6.23   *Circuit diagram for Practical Exercise 6.6*

Notice, for the 741 ic, that if the frequency is increased beyond about 15 kHz, there is a serious deterioration in the output voltage waveform.

3 Measure the time (t) for the output voltage to go from 10% to 90% of its final value (V_{max}).
4 Calculate the value of the slew rate.
5 Set the generator to give a 10 V peak-to-peak **sinusoidal** waveform at 10 kHz. The output voltage waveform at this point should also be sinusoidal.
6 Increase the frequency, noting the value (f_{max}) at which the output voltage waveform starts to become triangular.
7 Calculate the expected value of f_{max} and compare it with the observed value.
8 Replace the 741 with the LF351 ic and repeat procedures 2 to 7 inclusive.
   The pin connections of the LF351 are identical to those for the 741.
   You should find this part of the exercise particularly straightforward!

Conclusions

1 Compare the performance of the two ic's on the basis of (a) slew rate, (b) f_{max}.
2 Compare your calculated values with those from the manufacturer's data.
3 The 741 is a bipolar ic. What type is the LF351?

Differentiation and integration

As the reader will be aware these are mathematical terms which may recall happy (or unhappy) experiences. Fortunately, the practical applications of these processes are less daunting. The development of analog computers arose from the need to simulate the performance of physical systems (for example mechanical) by means of their electrical analog equivalent. Sufficient to say, integration rather more than differentiation networks played a large part in this process.

Our concern here is to use the two processes for **wave-shaping** purposes, by means of a practical exercise.

Practical Exercise 6.7

To investigate the action of a differentiating and integrating operational amplifier. For this exercise you will need the following components and equipment:

1 – ic operational amplifier (741)
2 – resistor (1 kΩ)
1 – capacitor (10 µF, 16 V)
1 – DC power supply (±15 V)
1 – audio frequency square wave generator
1 – double beam cathode ray oscilloscope

Procedure (a) The differentiator

1 Connect up the circuit shown in Figure 6.24(a). The pin connections for the 741 ic are given in Figure 6.5.
2 Set the input to a 1 V peak-to-peak square wave of frequency 100 Hz. Observe the input and output waveforms on the CRO.
3 Sketch both waveforms, which should be similar to those shown in Figure 6.25(a).

Procedure (b) The integrator

1 Connect up the circuit shown in Figure 6.24(b).
2 Set the input to a 1 V peak-to-peak square wave of frequency 100 Hz. Observe the input and output waveforms.
3 Sketch both waveforms, which should be similar to those shown in Figure 6.25(b).

Important points

For the differentiator (see Figure 6.25(a)):

- For a suitable square wave input, the output consists of a series of short duration pulses or 'spikes' which can be used to trigger or switch-on other circuits.
- It acts as a high pass filter, allowing high frequencies to pass and attenuating low frequencies.
- The time constant ($T = C_1 R_2$) should be small compared with the pulse width.

Continued on p. 156

Practical Exercise 6.7 (*Continued*)

Important points (*Continued*)

For the integrator (see Figure 6.25(b)):

- It acts as a low pass filter, removing high frequency components from a pulse waveform.
- If the time constant $(T = C_1 R_1)$ is large compared with the pulse width, the output will be triangular.
- It can be used to provide time delays in switching circuits.

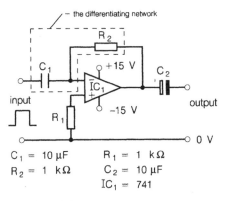

C_1 = 10 μF   R_1 = 1 kΩ
R_2 = 1 kΩ   C_2 = 10 μF
                 IC_1 = 741

(a) differentiating

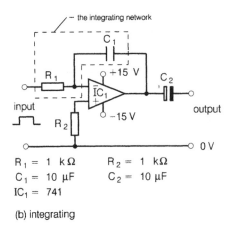

R_1 = 1 kΩ   R_2 = 1 kΩ
C_1 = 10 μF   C_2 = 10 μF
IC_1 = 741

(b) integrating

Figure 6.24 *The differentiating and integrating operational amplifier: circuit diagrams for Practical Exercise 6.7*

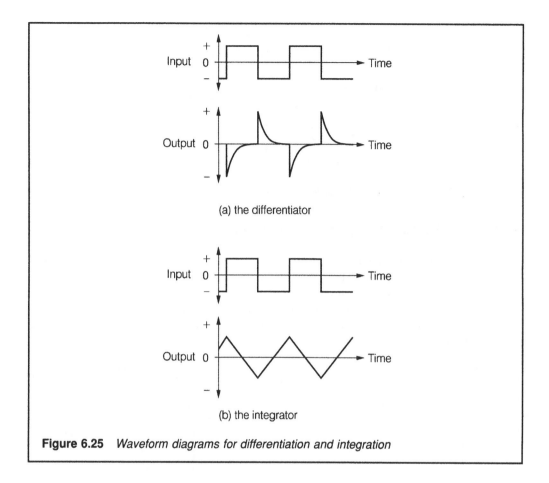

(a) the differentiator

(b) the integrator

Figure 6.25 *Waveform diagrams for differentiation and integration*

The use of single supply rails

The advantage in using a dual power supply for operational amplifiers is that the output is then able to swing either side of the zero-volt value. In addition, the output can be adjusted to zero when the differential input is zero. If there is no requirement for the output to vary either side of zero volts, then a single-polarity supply can be used. The necessary circuit modification for the inverting amplifier is shown in Figure 6.26.

Important points

- The non-inverting input terminal is held at a positive potential, usually one-half of the supply voltage, by the potential divider resistors R_3 and R_4.
- Capacitor C_1 in parallel with R_4 maintains the non-inverting input terminal as a short circuit to signal frequencies.

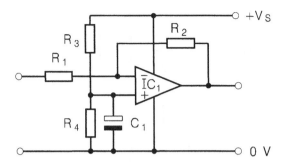

Figure 6.26   *The inverting operational amplifier: using a single power supply*

Field effect transistor operational amplifiers

For many years the 741 was the industry standard operational amplifier. It can still be regarded as a general purpose building block, able to perform the required functions at low financial cost. The development of field effect technology has inevitably created a number of FET operational amplifiers that can out-perform the 741 in most respects. The reader is invited to refer to manufacturer's data in order to complete the table given in Figure 6.27.

	741	LM358	LM837	LF351	TL071	TL082	CA3140
Type	bipolar						
Single/ dual/ quad	single						
Supply voltage	± 5 V to ± 18 V						
Power dissipation	500 mW						
Input impedance	$2\,M\Omega$						
Large signal voltage gain	106 dB						
Unity gain – bandwidth	1 MHz						
Slew rate	0.5 V/μs						
CMMR	90 dB						
Price							

Figure 6.27   *Table of data for integrated circuit operational amplifiers*

7 Oscillators

The concept of positive feedback

Consider a pendulum bob swinging to and fro. Figure 7.1 shows the arrangement. The bob swings through an arc from position 1 at height '*a*' to position 2 at height '*b*' and back. Energy is lost during each half-swing and the amplitude of that swing becomes progressively smaller than the one before. Its return swing is to position 3, then to position 4 and so on. The pendulum will eventually come to a standstill at its rest position.

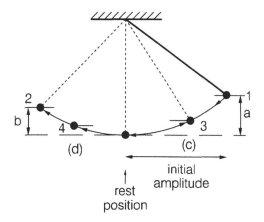

Figure 7.1 *The pendulum*

In order for the pendulum to continue to swing to the same height, that is, with constant amplitude, **two conditions** are necessary:

(1) The energy lost must be replaced **by the right amount**, that is, by the amount that was lost. If the replacement energy is too small the system will still come to a halt, and if too large it will go berserk, or in technical terms, become unstable.

(2) The replacement energy must be provided **at the right time**. This means giving the pendulum a push at the extreme of its travel, just as the movement is on the turn. If the push is given too soon it will oppose the swing and if too late it will be ineffective. The timing is therefore crucial, with the replacement energy having to

be provided **in phase** with the system movement. This in-phase requirement gives us the definition of positive feedback.

The pendulum example illustrates that the main condition for a continuous swing is the provision of the 'push', namely the energy or power source.

Now suppose that the movement of the pendulum bob caused a trace or waveform to be made on a piece of paper moving across the line of the bob at constant speed. The waveform would be as shown in Figure 7.2(a) and is known as a sinewave. The interesting fact about this is that the sinewave comes about as a result of something (*the pendulum swing*) happening **naturally**. Another example is that of a ruler being held at one end on a table while the other end is given a flick. The subsequent 'twang' of the movement dying away with time would again provide a sinewave trace, decreasing in amplitude as our original pendulum and giving a decaying sinewave, as shown in Figure 7.2(b). The swings of the pendulum and ruler are called oscillations, where the word oscillate means 'to move to and fro between two points'.

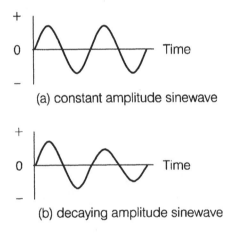

(a) constant amplitude sinewave

(b) decaying amplitude sinewave

Figure 7.2   *The sinewave and decaying sinewave*

One final point before we leave the pendulum. The number of swings in a given time (the frequency) is dependent upon the length of the pendulum, with a shorter length giving a higher frequency, and of course vice versa.

The oscillator

The electrical oscillator is a device which converts direct current into alternating current. The only input that an oscillator requires is from a DC power supply, which, with the correct conditions, will provide an alternating output from the oscillator, see Figure 7.3(a). An alternative description for the oscillator is that of an amplifier providing its own input with the use of **positive** feedback being all-important. See Figure 7.3(b).

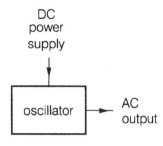

(a) basic block diagram

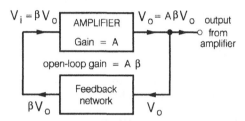

(b) amplifier providing its own input

Figure 7.3 *The oscillator*

Readers will recall that negative feedback, in which a fraction of the output was fed back out of phase to the input, gave an overall gain of A', where

$$A' = \frac{A}{1 + A\beta}$$

Positive feedback, in which a fraction of the output is fed back **in phase** to the input, will give an overall gain of A', where

$$A' = \frac{A}{1 - A\beta}$$

As before, the term $A\beta$ is called the open-loop gain.

Electrical oscillators can be designed to provide a range of waveforms, namely sinusoidal, rectangular, sawtooth and triangular.

Important points

- For oscillations to start:
    (i) The loop gain $A\beta = 1$, giving amplifier gain $A = 1/\beta$. With this arrangement the overall gain A' will be infinity.
    (ii) The overall phase shift round the loop must be zero ($= 360°$).

- Oscillators have **two** main sections:
    (i) A maintaining or amplifier section, to replace the losses.

Continued on p. 162

Important points (*Continued*)

(ii) A frequency determining section, which can be either an *RC* or *LC* network.

- Amplifier phase shift + feedback network phase shift $= 0° = 360°$

Sinusoidal oscillators

Examples of this type of oscillator are found in signal generators, touch-tone telephones, musical instruments and radio and television transmitters.

The 3RC phase-shift oscillator

This type of sinusoidal oscillator uses a $3R$ and $3C$ network as the feedback circuit and a practical exercise will consider an example of this network.

Practical Exercise 7.1

To investigate the *RC* phase-shift network.
For this exercise you will need the following components and equipment:

3 – resistor (4.7 kΩ)
3 – capacitor (0.01 µF)
1 – audio frequency signal generator
1 – double beam cathode ray oscilloscope

Procedure (a)

1 Connect up the circuit of Figure 7.4.

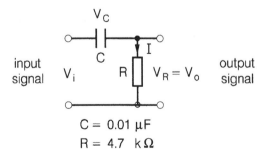

$C = 0.01 \ \mu F$
$R = 4.7 \ k\Omega$

Figure 7.4 *The single RC phase-shift network: circuit diagram for Practical Exercise 7.1 (Procedure (a))*

2 Set the signal generator frequency to 5 kHz and its output voltage to 1 V peak-to-peak. Display both input and output waveforms of the network on the CRO. Measure the output voltage and its phase shift with respect to the input voltage, stating whether the former is leading or lagging.

For those readers who are unfamiliar with or just a little rusty on measuring phase shift with a CRO, refer to the diagram in Figure 7.5.

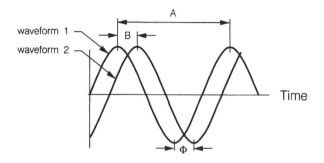

* waveform 1 LEADS waveform 2 by angle Φ,

where $\Phi = \dfrac{B}{A} \times 360$ degrees

* measure A and B in divisions

Figure 7.5 *Measurement of phase shift with CRO*

For those who are familiar with phasor diagrams, an example for this circuit is given in Figure 7.6.

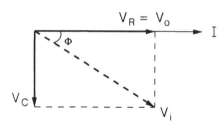

* output voltage V_O LEADS input voltage V_i by angle Φ

* $\cos \Phi = \dfrac{V_O}{V_i} = \dfrac{I_R}{I_Z} = \dfrac{R}{Z}$

* $\tan \Phi = \dfrac{V_C}{V_O} = \dfrac{I_{X_C}}{I_R} = \dfrac{1}{2\pi fCR}$

Figure 7.6 *Phasor diagram for the RC phase-lead network*

Continued on p. 164

Practical Exercise 7.1 (*Continued*)

3 Repeat procedure 2 at frequencies of 4 kHz, 3 kHz, 2 kHz and 1 kHz.
4 Complete the table in Figure 7.7, which also provides the theoretical answers.

f (kHz)	Measured Φ (deg)	Measured $\dfrac{V_o}{V_i}$	Calculated Φ (deg)	Calculated $\dfrac{V_o}{V_i}$
1			73.5	0.28
2			59.4	0.51
3			48.5	0.66
4			40.2	0.76
5			34.1	0.83

Figure 7.7   *Table of results for Practical Exercise 7.1*

Important points

For the single *RC* network:

- The maximum phase shift between output and input voltage approaches 90°, as the frequency approaches 0 Hz.

- The output voltage is always less than the input voltage, since this is a passive network (there being no amplification).

- For the phase shift to be 90° at a particular frequency the value of *R* would have to be zero, at which point the output voltage would also be zero! Hence to obtain an overall phase shift of 180°, at least three similar *RC* networks are needed. The arrangement is called a ladder network.

Procedure (b)

1 Connect up the circuit of Figure 7.8.
2 With an input signal voltage of 10 V peak-to-peak (the actual value is not critical) display both input and output signals on the CRO.
3 Vary the frequency of the signal generator over the range 500 Hz to 1500 Hz and note the frequency at which the phase difference between the two waveforms is 180°.

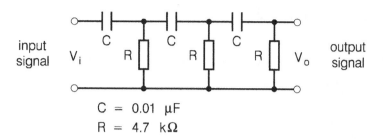

$$C = 0.01 \ \mu F$$
$$R = 4.7 \ k\Omega$$

Figure 7.8 *The 3RC phase-shift network: circuit diagram for Practical Exercise 7.1 (Procedure (b))*

Estimate the value of the voltage ratio v_o/v_i.

Important points

For the 3RC network:

- The frequency (f) at which the phase shift is 180° can be calculated from

$$f = \frac{1}{2\pi RC \sqrt{6}} \text{ hertz (Hz)}$$

with R in ohms (Ω) and C in farads (F).
    The voltage ratio v_o/v_i at which the phase shift is 180° is 1/29. In other words, the attenuation is 29.

- It cannot be assumed that the contributions of each RC branch to the overall 180° phase shift are equal.

- The loss in the feedback network is our old friend, the feedback fraction β, and so, in this case, $\beta = 1/29$.

For the 3RC oscillator:

- The gain of the amplifying section must be at least equal to the loss or attenuation in the feedback network. In practice the gain will need to be larger than 29 in order to **sustain** the oscillations.

Conclusions

1 Compare theoretical and measured values of frequency that give 180° phase shift.
2 Why might the phase shift for each part of the network not all be 60°?

Practical Exercise 7.2

To investigate the action of a 3RC phase-shift transistor oscillator.
  For this exercise you will need the following components and equipment:

1 – npn transistor (BC109)
3 – resistor (4.7 kΩ)
1 – resistor (10 kΩ, 47 kΩ)
1 – potentiometer (1 kΩ)
3 – capacitor (0.01 μF)
2 – capacitor (100 μF, 16 V)
1 – DC power supply (+12 V)
1 – double beam cathode ray oscilloscope

Procedure

1 Connect up the circuit shown in Figure 7.9. The pin connection diagram for the BC109 transistor is shown in Figure 3.5.

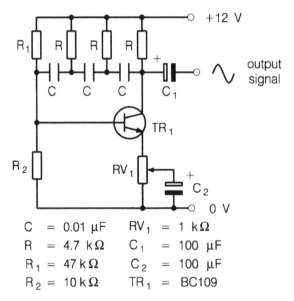

C = 0.01 μF	RV$_1$ = 1 kΩ
R = 4.7 kΩ	C$_1$ = 100 μF
R$_1$ = 47 kΩ	C$_2$ = 100 μF
R$_2$ = 10 kΩ	TR$_1$ = BC109

Figure 7.9 *The 3RC transistor phase-shift oscillator: circuit diagram for Practical Exercise 7.2*

2 It may be necessary to adjust resistor R_3 in order to achieve oscillations. This adjustment will vary the gain of the amplifier and should be in the position that gives a reasonable sinusoidal waveform.

Important point

- A distorted output waveform is the result of too high an amplifier gain – remember the pendulum swing where the energy replacement was too great?

---

3 Measure the frequency of the oscillation and compare with the calculated value, given by

$$f = \frac{1}{2\pi RC\sqrt{6}} \text{ Hz, where } R \text{ and } C \text{ are the components of the } 3RC \text{ network.}$$

Conclusion

1 The difference between the measured and calculated values is usually due to a so-called 'loading' effect. What do you think this refers to?

The 2RC phase-shift oscillator

This very useful and important network, called a Wien Bridge, is most often used with an operational amplifier.

Practical Exercise 7.3

To investigate the $2RC$ phase-shift Wien Bridge network.
  For this exercise you will need the following components and equipment:

2 – resistor (4.7 kΩ)
2 – capacitor (0.01 µF)
1 – audio frequency signal generator
1 – double beam cathode ray oscilloscope

Procedure

1 Connect up the circuit shown in Figure 7.10.
2 Set the signal generator output voltage to 10 V peak-to-peak and display both input and output waveforms of the network on the CRO.
3 Vary the frequency of the signal generator over the range 2 kHz to 5 kHz and note the frequency at which the phase difference between the two waveforms is

---

Continued on p. 168

Practical Exercise 7.3 *(Continued)*

180°. At this frequency,

estimate the value of the voltage ratio $\dfrac{V_o}{V_i}$

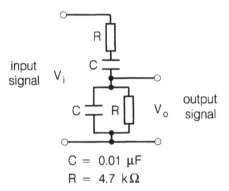

$$C = 0.01\ \mu F$$
$$R = 4.7\ k\Omega$$

Figure 7.10   *The 2RC phase-shift network: circuit diagram for Practical Exercise 7.3*

Important points

For the Wien Bridge network:

- The phase shift between the output and input voltage will be zero at a frequency (f) given by

$$f = \frac{1}{2\pi RC} \text{ hertz (Hz)}$$

where R is in ohms (Ω) and C in farads (F).

- At the frequency where the phase shift is zero, the attenuation is 3, that is, $v_o/v_i = 1/3$.

Practical Exercise 7.4

To investigate the action of an integrated circuit Wien Bridge oscillator.
For this exercise you will need the following components and equipment:

1 – ic operational amplifier (741)
2 – resistor (4.7 kΩ)
1 – resistor (1 kΩ, 1.8 kΩ)

1 – variable resistor (4.7 kΩ)
2 – capacitor (0.01 μF)
1 – capacitor (100 μF, 16 V)
2 – diode (1N4148)
1 – DC power supply (±15 V)
1 – double beam cathode ray oscilloscope

Procedure

1 Connect up the circuit shown in Figure 7.11(a). The pin connection diagram for the 741 ic is given in Figure 6.5.

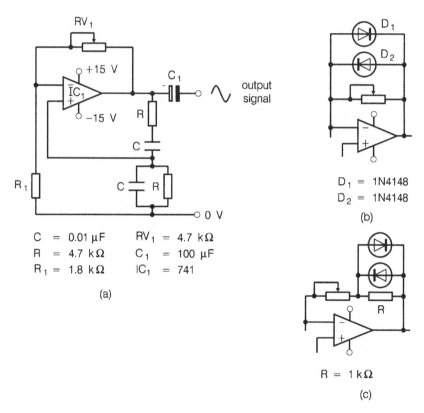

C = 0.01 μF RV_1 = 4.7 kΩ
R = 4.7 kΩ C_1 = 100 μF
R_1 = 1.8 kΩ IC_1 = 741

(a)

D_1 = 1N4148
D_2 = 1N4148

(b)

R = 1 kΩ

(c)

Figure 7.11 *The integrated circuit Wien Bridge oscillator: circuit diagrams for Practical Exercise 7.4*

2 Vary the value of RV_1 from minimum to maximum (thus giving minimum to maximum amplifier gain) and notice that the corresponding effects are:

(a) no oscillation and therefore no output
(b) the output waveform is very distorted.

Continued on p. 170

Practical Exercise 7.4 (*Continued*)

3 Adjust the value of RV_1 until a reasonably undistorted sinewave is shown on the CRO.

4 Adjust RV_1 to give the maximum undistorted output voltage and notice that the peak value is just less than the saturated voltage level of the operational amplifier.

Important point

- The adjustment of amplifier gain for satisfactory operation is quite critical and the output waveform can be improved by limiting the amplitude of the oscillations.

---

5 Connect the two diodes as shown in Figure 7.11(b). Note that they are connected in opposite directions, called inverse parallel.

6 Adjust RV_1 to give maximum undistorted output and note that the peak-to-peak value of the output waveform is now approximately 1.2 V, which equals twice the diode forward voltage-drop.

Important points

- Each diode will conduct when the forward voltage across it exceeds approximately 0.6 V. The forward-conducting diodes have very low resistance and being in parallel with the feedback resistor, will cause the overall voltage gain of the op-amp to fall.

- The oscillator can be varied in frequency if either the two capacitors (*C*) or the two resistors (*R*) are variable. It is more convenient to vary the resistors (together) since dual ganged types are readily available.

---

7 Modify the circuit to that shown in Figure 7.11(c). You should find that although the amplitude is not limited as in procedure 6 (the amplifier could saturate) the control over obtaining the sinusoidal output is easier.

Conclusion

1 The frequency of the circuit in Figure 7.11(a) is to be varied between 1 kHz and 5 kHz. The resistors *R* consist of a fixed resistor in series with a variable resistor. The capacitors *C* are both 0.01 μF.

   Calculate the values of the fixed and variable resistors required. Use the manufacturer's catalogue to choose the most appropriate preferred values.

Applications for the RC oscillator

The *RC* network is used for oscillators operating over the frequency range from 1 Hz to about 1 MHz. In particular, the Wien Bridge is widely used for laboratory oscillators and provides a stable signal output with low distortion.

The LC tuned circuit oscillator

A tuned circuit consists of an inductor *L* in parallel with a capacitor *C*. The electrical principles tell us that providing the resistance of the inductor is negligible, the resonant frequency (f_r) of this parallel combination is given by

$$f_r = \frac{1}{2\pi\sqrt{LC}} \text{hertz (Hz)}$$

where *L* is in henrys (H) and *C* in farads (F).

At this frequency, the voltage across the coil will be a maximum. The tuned circuit is used as an alternative to the resistor as the load in the collector circuit of a transistor amplifier, with the voltage gain of this amplifier thus being a maximum at the resonant frequency.

A secondary coil placed close to the main (primary) coil (*L*) can be connected such that an in-phase fraction of the output across the coil *L* is fed back to the input.

The principal requirements for oscillation remain the same as previously described.

The basic circuit of the tuned circuit transistor oscillator is shown in Figure 7.12. For simplicity, no bias arrangements are shown.

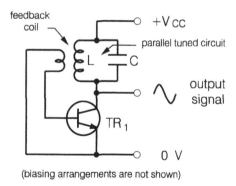

(biasing arrangements are not shown)

Figure 7.12 *The transistor tuned circuit LC oscillator: the basic circuit*

The tuned circuit of *L* and *C* in parallel provide oscillations at a frequency (f_o) given by

$$f_o = \frac{1}{2\pi\sqrt{LC}} \text{ hertz (Hz)}$$

where *L* is in henrys (H) and *C* in farads (F).

A practical *LC* oscillator circuit can consist of either a tapped coil (Hartley) or a tapped capacitor (Colpitts). Since a tapped capacitor arrangement is easier to set up, this type will be used in the practical exercise which follows.

Practical Exercise 7.5

To investigate the action of a transistor *LC* Colpitts oscillator.
  For this exercise you will need the following components and equipment:

  1 – npn transistor (BC109)
  1 – inductor (100 µH)
  1 – capacitor (47 nF, 100 nF)
  1 – capacitor (0.01 µF, 0.47 µF)
  1 – resistor (10 kΩ, 47 kΩ)
  1 – variable resistor (1 kΩ)
  1 – DC power supply (+12 V)
  1 – double beam cathode ray oscilloscope

Procedure

1 Connect up the circuit shown in Figure 7.13. The pin connection diagram for the BC109 transistor is shown in Figure 3.5.

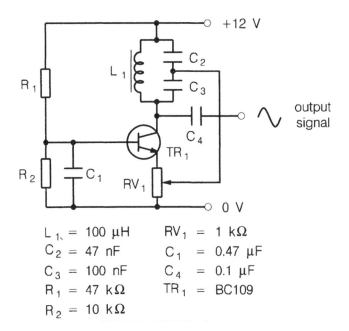

$$L_1 = 100 \ \mu H \qquad RV_1 = 1 \ k\Omega$$
$$C_2 = 47 \ nF \qquad C_1 = 0.47 \ \mu F$$
$$C_3 = 100 \ nF \qquad C_4 = 0.1 \ \mu F$$
$$R_1 = 47 \ k\Omega \qquad TR_1 = BC109$$
$$R_2 = 10 \ k\Omega$$

Figure 7.13 *The bipolar transistor Colpitts oscillator: circuit diagram for Practical Exercise 7.5*

2 Connect the CRO to the output and adjust RV_1 for a suitable amplitude of oscillations.

3 Measure the frequency of the oscillations and compare this with the calculated value. Note that the value of C in the above expression for f_0 is the result of two capacitors in series, namely

$$C = \frac{C_1 C_2}{C_1 + C_2}$$

4 Change over capacitors C_2 and C_3 and note any differences (if any) in the frequency and the amplitude of the oscillations.

Conclusion

1 Explain fully the purpose of the variable resistor RV_1.
  Hint. You may find it helpful to refer back to Chapter 5.

Practical Exercise 7.6

To investigate the action of an integrated circuit *LC* Colpitts oscillator.
  For this exercise you will need the following components and equipment:

  1 – ic operational amplifier (741, LF351)
  1 – inductor (100 µH)
  1 – capacitor (47 nF, 100 nF)
  1 – capacitor (0.01 µF)
  1 – resistor (8.2 kΩ, 10 kΩ, 47 kΩ)
  1 – DC power supply (±15 V)
  1 – double beam cathode ray oscilloscope

Procedure

1 Connect up the circuit shown in Figure 7.14. The pin connection diagram for the 741 integrated circuit is shown in Figure 6.5.
2 Connect the CRO to observe the output waveform.
3 It is quite likely that the circuit will not oscillate due to the restrictions imposed by the slew rate and bandwidth of the 741.
4 Replace the 741 ic with the LF351 ic. (The pin connections for the LF351 and 741 integrated circuits are identical.)
5 The circuit should now oscillate. Measure the frequency of the oscillations.
6 Calculate the theoretical value of the frequency (f) from

$$f = \frac{1}{2\pi\sqrt{LC}}, \quad \text{where } C = \frac{C_1 C_2}{C_1 + C_2}$$

Continued on p. 174

Practical Exercise 7.6 (*Continued*)

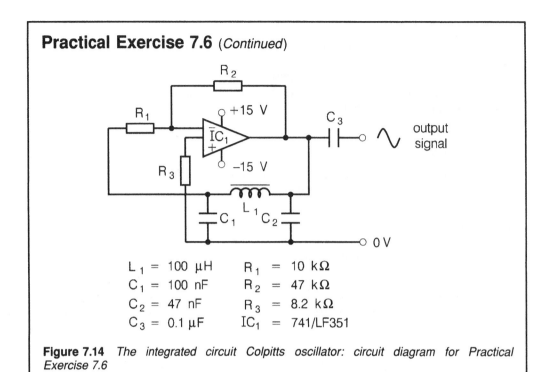

$$L_1 = 100 \ \mu H \qquad R_1 = 10 \ k\Omega$$
$$C_1 = 100 \ nF \qquad R_2 = 47 \ k\Omega$$
$$C_2 = 47 \ nF \qquad R_3 = 8.2 \ k\Omega$$
$$C_3 = 0.1 \ \mu F \qquad IC_1 = 741/LF351$$

Figure 7.14 *The integrated circuit Colpitts oscillator: circuit diagram for Practical Exercise 7.6*

Application for the LC oscillator

This type of oscillator is used for radio frequencies in the MHz region. To use this oscillator for frequencies less than 1 MHz would entail large physical sizes for both L and C. The RC oscillator is therefore more suitable for this purpose.

Non-sinusoidal oscillators

These are oscillators which rely on the charging and discharging of a capacitor through a resistor. They are called **relaxation** oscillators and their output waveform is sawtooth or rectangular. The active device (that is, the transistor or integrated circuit) is switched on and off periodically.

The astable multivibrator

An important example of the relaxation oscillator is the astable **multivibrator** whose ideal waveform is shown in Figure 7.15.

The word **astable** means not stable, or free-running, thus giving a continuous train of rectangular pulses. The word multivibrator arises from the fact that rectangular (and square) waveforms contain many harmonics of the fundamental frequency. Another name for the multivibrator oscillator is 'flip-flop'.

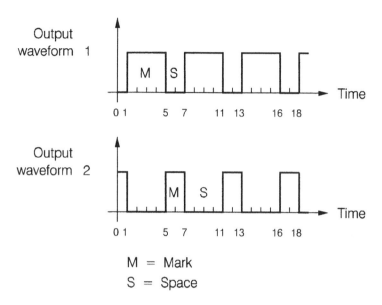

Figure 7.15   *The astable multivibrator: typical output waveforms*

Important points

- The **mark** (M) is the time for the positive part of the waveform, called the 'on' time.

- The **space** (S) is the time for which there is no waveform, or for which it is negative, called the 'off' time.

- The mark–space ratio (M/S) is therefore given by

$$\frac{M}{S} = \frac{\text{the 'on' time}}{\text{the 'off' time}}$$

- For the waveforms shown:

  Waveform 1 shows $M = 4$ (time units) and $S = 2$, giving a mark–space ratio of 4/2, or 2:1
  Waveform 2 shows $M = 2$ and $S = 4$, which gives a mark-space ratio of 2/4, or 1:2

The circuit for an astable multivibrator using two bipolar transistors connected in common emitter is shown in Figure 7.16.

The output of each transistor is fed back to the input of the other transistor.

Since each (common emitter) amplifier provides 180° phase shift, there is a total of 360° for the closed loop, thus giving the conditions for oscillation.

Typical output waveforms are shown in Figure 7.17.

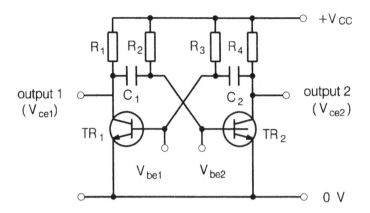

Figure 7.16   *The astable multivibrator: basic circuit*

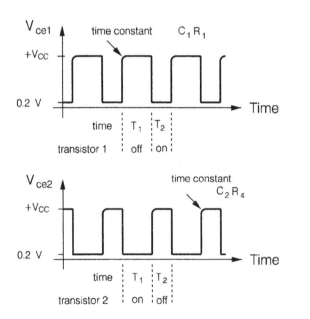

Figure 7.17   *The astable multivibrator: Waveform diagrams relating to the circuit of Figure 7.16*

Important points

For the waveforms of Figure 7.17 relating to the circuit of Figure 7.16:

- Output from transistor 1 (V_{ce1}):

  Calculate the mark, T_1, as $T_1 = 0.7C_2R_3$
  Calculate the space, T_2, as $T_2 = 0.7C_1R_2$

- Output 2 from transistor 2:

  Calculate the mark, T_2, as $T_2 = 0.7C_1R_2$
  Calculate the space, T_1, as $T_1 = 0.7C_2R_3$

- T_1 and T_2 are in seconds with C_1 and C_2 in farads and R_1 and R_2 in ohms.

- The frequency (f) of the output waveform, usually referred to as the **pulse repetition frequency** (prf), is given by

$$f = \frac{1}{\text{time for one complete cycle}}$$

$$= \frac{1}{T_1 + T_2}$$

$$= \frac{1}{0.7(C_2R_3 + C_1R_2)} \text{ hertz}$$

for C_1 and C_2 in farads and R_2 and R_3 in ohms.

- If $C_1 = C_2 = C$ and $R_2 = R_3 = R$, then the mark–space ratio for each output is 1, with the resulting waveforms being square. In this case, the frequency (f) is given by

$$f = \frac{1}{1.4CR} \text{ hertz (Hz)}$$

Practical Exercise 7.7

To investigate the action of a transistor astable multivibrator (1).
  For this exercise you will need the following components and equipment:

2 – npn transistor (BC109)
2 – resistor (270 Ω, 15 kΩ)
2 – capacitor (47 μF, 16 V)
2 – LED (5 mm)
1 – DC power supply (+12 V)

This 'flashing light' exercise will illustrate the switching action of the astable multivibrator.

Procedure

1 Connect up the circuit shown in Figure 7.18. The pin connection for the BC109 transistor is given in Figure 3.5.
2 Confirm that the two LEDs flash on and off alternately at a frequency of approximately 1 Hz. Remember that one cycle of operation corresponds to one 'on/off' period for **one** LED, or alternatively, one 'on' period for **each** LED.

Continued on p. 178

Practical Exercise 7.7 (*Continued*)

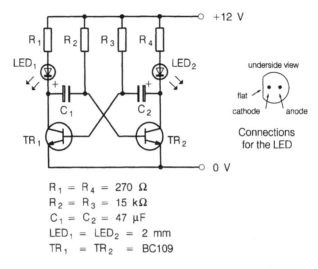

$R_1 = R_4 = 270 \ \Omega$
$R_2 = R_3 = 15 \ k\Omega$
$C_1 = C_2 = 47 \ \mu F$
$LED_1 = LED_2 = 2 \ mm$
$TR_1 = TR_2 = BC109$

Figure 7.18 *The astable multivibrator oscillator: circuit diagram for Practical Exercise 7.7*

3 Change the value of capacitor C_1 to 100 μF and notice that:
   (a) LED_1 remains lit for a longer period of time than LED_2
   (b) the frequency is now less than in Procedure 2 above.

Important point

• For the astable multivibrator only **one** transistor can be on at any particular time, during which the other transistor is off.

Practical Exercise 7.8

To investigate the action of a transistor astable multivibrator (2).
   For this exercise you will need the following components and equipment:

2 – npn transistor (BC109)
2 – resistor (1 kΩ)
2 – resistor (33 kΩ)
1 – capacitor (0.01 μF)
2 – capacitor (0.02 μF)
1 – DC power supply (+12 V)
1 – double beam cathode ray oscilloscope

Procedure

1 Connect up the circuit shown in Figure 7.19. The pin connection diagram for the BC109 transistor is given in Figure 3.5.

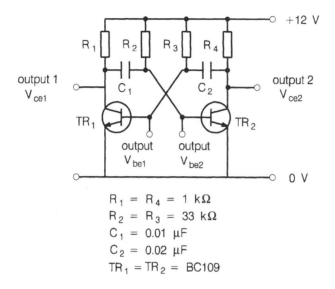

$$R_1 = R_4 = 1 \text{ k}\Omega$$
$$R_2 = R_3 = 33 \text{ k}\Omega$$
$$C_1 = 0.01 \ \mu\text{F}$$
$$C_2 = 0.02 \ \mu\text{F}$$
$$TR_1 = TR_2 = \text{BC109}$$

Figure 7.19 *The astable multivibrator: circuit diagram for Practical Exercise 7.8*

2 Display the output waveforms from the collector of each transistor and measure the times for the mark and space for each of them.
3 Using the results from Procedure 2, calculate for each waveform:
   (a) the mark–space ratio and
   (b) the frequency.
4 Change C_1 to 0.02 μF and repeat Procedures 2 and 3.
5 Complete the table shown in Figure 7.20.
6 Display the waveforms at the base and collector of transistor 1. Sketch these waveforms on separate axes but to the same time scale. They should resemble those shown in Figure 7.21.
   Make sure that:
   (i) the height, in volts, of each waveform is correctly marked
   (ii) the waveforms are at the correct DC level – for this to be so, the traces on the CRO must be 'zeroed'.

Conclusions

1 Compare the measured values of frequency for the respective values of C_1 and C_2, with the calculated values.
2 From the waveform diagrams drawn in procedure 6, estimate the value of:

Continued on p. 180

Practical Exercise 7.8 *(Continued)*

		$C_1 = 0.01$ µF $C_2 = 0.02$ µF		$C_1 = C_2 = 0.02$ µF	
		V_{ce1}	V_{ce2}	V_{ce1}	V_{ce2}
Measure	time for mark				
	time for space				
Calculate	mark–space ratio				
	frequency(f)				
	Calculate ⟶	$f = \dfrac{1}{1.4CR}$		$f = \dfrac{1}{1.4CR}$	

Figure 7.20   *Table of results for Practical Exercise 7.8*

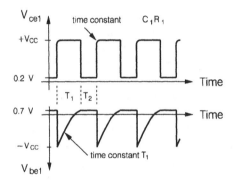

Figure 7.21   *The astable multivibrator: waveform diagrams for Practical Exercise 7.8*

(a) the base voltage and
(b) the collector voltage, both when TR_1 is fully conducting.

Important points

- When one transistor is on, or conducting ($V_{CE} = +0.2$ V), the other is off, or not conducting ($V_{CE} = +12$ V).

- The astable multivibrator resembles a see-saw where one end is up in the air while the other is down on the ground. Neither end can be in the same position at the same time.

- The change of state (transistor switching from on to off and vice versa) happens very quickly, hence the vertical edges of the collector waveforms.

Application for the astable multivibrator

An astable multivibrator is used for precision timing applications, as required in computer equipment, which includes computers themselves and modern keyboard instruments and their associated devices. The need is for a series, or 'train' of rectangular clock pulses. For these applications, various types of integrated circuits are generally used. If further details on this aspect are needed, they can be found in the author's previous book: *Practical Digital Electronics for Technicians*, also from Butterworth-Heinemann.

Questions

7.1 Explain the difference between negative and positive feedback.

7.2 Justify the following statement: '*An oscillator is an amplifier that provides its own input.*'

7.3 What is the greatest phase shift (almost) possible with a combination of a capacitor and a resistor?

7.4 A certain *RC* phase-shift oscillator has a frequency of oscillation given by:

$$f = \frac{1}{2\pi\sqrt{6}RC},$$

with all quantities in their basic units

(a) $R = 4.7$ kΩ. What value of C is required to give a frequency of 1 kHz?

(b) $C = 47$ nF. What value of R is required for a frequency of 5 kHz?

7.5 A Wien Bridge network is required to give zero phase shift at a frequency of 15 kHz. If the capacitors used are each 1000 pF, calculate the necessary value for each resistor. What minimum amplifier gain is needed when using this network to provide oscillations?

7.6 An *LC* tuned circuit oscillator has $L = 100$ μH and $C = 100$ pF. At what theoretical frequency should it oscillate?

If L remains constant at 100 μH, what variation in C will cause a frequency variation of 10:1?

7.7 A transistor astable multivibrator is required to have a natural frequency of 10 kHz with unity mark-to-space ratio. The base resistors each have a value of 180 kΩ. Calculate the necessary value of the coupling capacitors.

8 Test and measuring equipment

A description of the essential items of equipment was given in Chapter 1 and the purpose of this chapter is twofold:

(a) To outline the general specification of test equipment used in performing the general tasks in an electronic environment at this level of study.
(b) To draw attention to the limitations and possible sources of error in measurement with the use of this equipment.

It is fair to assume that the following items of equipment are available:

(a) Voltmeter, ammeter and ohmmeter.
   The ammeter is not strictly essential as a separate item, since in most cases, a measurement of voltage followed by the use of Ohm's Law would be a helpful alternative.
   In practice these three features are more likely be combined in the form of the multimeter.
(b) Signal generator covering the frequency 10 Hz to 50 kHz+.
(c) Cathode ray oscilloscope, preferably double beam.

General considerations

The purpose of test equipment is to provide accurate working data for a particular circuit and it is therefore necessary to make sure that the procedures in use perform this task, with the **minimum of disturbance** to the circuit in question. As the following paragraphs will highlight, this is not always as easy and clear-cut as it may seem.
   Taking each item in turn:

The voltmeter

Since a voltmeter will measure voltage **across** a component, it must be placed in parallel with that component, as shown in Figure 8.1.
   A voltmeter is essentially a high resistance device (ideally infinite) whose parallel position in the circuit should not affect the operation of the circuit. In reality, depending

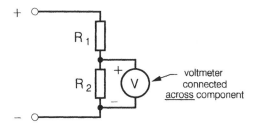

Figure 8.1 *Measurement of voltage*

on the type of voltmeter, the actual operating resistance can range from a few kΩ to a few MΩ. Whether or not this results in a detrimental effect on the circuit operation depends upon the resistance across which the voltmeter is placed. This paralleling of the voltmeter is called 'loading'. We will use the first practical exercise of this chapter to illustrate this effect.

Practical Exercise 8.1

To investigate the effect on circuit operation of the use of test equipment.
For this exercise you will need the following components and equipment:

1 – DC voltmeter (analog)
1 – DC voltmeter (digital)
2 – resistor (1 kΩ, 10 kΩ, 100 kΩ)
1 – DC power supply (+12 V)

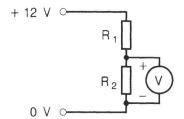

Figure 8.2 *Circuit loading: circuit diagram for Practical Exercise 8.1*

Procedure

1 Connect up the circuit shown in Figure 8.2, with $R_1 = R_2 = 1$ kΩ.
2 Measure the voltage across R_2:

    (a) with the analog meter on the 10 V range
    (b) with the analog meter on the 100 V range
    (c) with the digital meter using the lowest possible range.

Continued on p. 184

Practical Exercise 8.1 (*Continued*)

3 Repeat procedure 2 for:

(a) $R_1 = R_2 = 10 \text{ k}\Omega$
(b) $R_1 = R_2 = 100 \text{ k}\Omega$

4 Complete the table of results shown in Figure 8.3.

$R_1 = R_2 \longrightarrow$	1 kΩ	10 kΩ	100 kΩ
Meter ⌐	V	V	V
Analog 10 V range			
Analog 100 V range			
Digital			

Figure 8.3   *Table of results for Practical Exercise 8.1*

Conclusions

1 Comment on the measured values of the voltage across R_2, as R_1 and R_2 increase in value.
2 What is the **general** effect of connecting an analog meter across a very high resistance circuit?
3 Does the use of a higher voltage range on the analog meter **necessarily** give a more accurate reading?
4 Does the digital multimeter give answers closer to those expected?

Important points

- The analog DC meter is a moving coil meter which requires a current through the coil in order for the pointer to move across the scale. This current must be provided **from the circuit under test**.

- The current for a particular voltage indication **decreases** as the voltage range selected increases.

- The digital meter takes very little current from the test circuit.

Let us examine this situation more closely. For the circuit shown in Figure 8.2 we need to measure the voltage across resistor R_2. At this point the reader may correctly suggest that a voltmeter is hardly necessary since the expectation is that the voltage across R_2 will be one-half of the supply voltage of $+12$ V, namely 6 V!

Nevertheless, the results obtained above will have shown that this expectation will not always be realized, which is the purpose of the calculations which eventually follow.

The analog DC meter requires a certain current for full-scale deflection (FSD), with this current being the same value **no matter what voltage range is chosen**. A typical current for FSD is 50 μA, being the most likely value if a multimeter was used in the above practical exercise.

Suppose we choose the 10 V range. The current of 50 μA for a deflection of 10 V corresponds to a meter resistance of

$$\frac{10 \text{ V}}{50 \text{ μA}}, \text{ or } 200 \text{ k}\Omega$$

Taking now the 100 V range, the current of 50 μA for a deflection of 100 V shows that the meter resistance is now

$$\frac{100 \text{ V}}{50 \text{ μA}}, \text{ or } 2 \text{ M}\Omega$$

Looking at it a different way we can say that:

(a) the resistance of 200 kΩ for the 10 V range gives a value of 200 kΩ/10 V or 20 kΩ/V (ohms per volt),
(b) the resistance of 2 MΩ for the 100 V range gives a value of 2 MΩ/100 V, which again equals 20 kΩ/V,
(c) the reciprocal, or inverse, of the ohms per volt value gives 1 V/20 kΩ or 50 μA which is the current given earlier for full-scale deflection.

Important points

- The current required for full-scale deflection is the **sensitivity** of the instrument.

- The alternative definition of sensitivity is the figure of ohms per volt (Ω/V).

- The resistance of a meter on a particular voltage range is found by multiplying the ohms per volt value by that range value.

Looking now at Figure 8.4(a), the measurement taken with $R_1 = R_2 = 1 \text{ k}\Omega$.

On the 10 V range, the voltmeter resistance (R_m) is 200 kΩ (much larger than R_2) and the circuit can be redrawn as shown in Figure 8.4(b) where the voltmeter is shown as a resistor.

The effective resistance (R') of R_2 in parallel with R_m is given by

$$R' = \frac{R_2 \times R_m}{R_2 + R_m} \quad \frac{\text{(product)}}{\text{(sum)}}$$

$$= \frac{200 \text{ k} \times 1 \text{ k}}{201 \text{ k}} = 0.995 \text{ k}\Omega$$

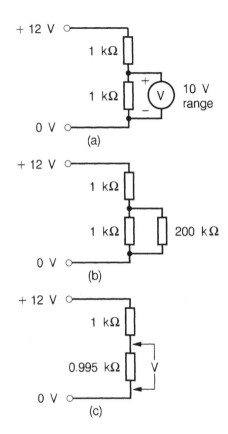

Figure 8.4   *Circuit diagrams to illustrate circuit loading*

The circuit can now be redrawn again, as in Figure 8.4(c), from which the value of the voltage across the parallel combination (V) can be calculated as

$$V = \frac{0.995 \text{ k}}{1.995 \text{ k}} \times 12 \text{ V}$$
$$= 5.98 \text{ V}$$

No real problem here since the calculated value with the meter present is close to that expected without.

Similar calculations for the other set of resistance values will give:

(a) for $R_1 = R_2 = 10$ kΩ, $V = 5.85$ V
(b) for $R_1 = R_2 = 100$ kΩ, $V = 4.80$ V

These **are not** the expected values of the voltage across resistor R_2. It is important to note, however, that they **are** the **correct** values **with the voltmeter connected**. This represents the **loading effect** of the voltmeter.

The resistance of the meter on the 100 V range will be 20 kΩ/V $\times$ 100 V, or 2 MΩ. Calculations for the actual voltages across R_2 when this range is used, will give 5.99 V, 5.97 V and 5.85 V respectively.

The problem using the higher voltage range is confined to the highest value of R_1 and R_2.

The measurement of AC voltages must not be forgotten. For this purpose, the analog instrument is most likely to be the moving-coil-plus-rectifier type. The sensitivity will be much smaller, typically 5 kΩ/V. Since measurements of AC voltage are normally confined to the low-impedance mains and the like, the loading effect is negligible. The meter will be calibrated for sinusoidal waveforms and any attempt to measure 'signal' voltages must keep this fact in mind.

Accuracy of readings

The analog meter used on a higher voltage range will offer a higher resistance to the circuit under test. The loading effect will be reduced but the deflection of the pointer will be smaller. This raises the question of meter accuracy and the ability of the operator to interpret these smaller deflections which themselves represent a larger voltage value on a higher voltage range.

The specification for meter accuracy is quoted as a percentage of the full-scale reading, typically ±1% on DC voltage ranges.

Thus a reading of 10 V on the 10 V range could actually have a value between 9.9 V and 10.1 V, whereas the same 10 V on the 100 V range could be between 9 V and 11 V. To this must be added the 'human error' in reading the smaller deflections.

The accuracy on AC voltage ranges is typically ±2.5% of full-scale.

The digital voltmeter has a constant input resistance, typically 10 MΩ, irrespective of the chosen range. The lowest necessary range can therefore be chosen with no adverse effect. Many digital meters are 'autoranging' in which no choice of range is possible or necessary. Typical accuracy is ±1% of the reading.

Important points

In general:

- The larger the resistance of the voltmeter compared to the resistance across which it is placed, the smaller the loading effect will be.

For analog voltmeters:

- Using a higher voltage range reduces the loading (good) but gives a smaller deflection (not so good).

- Always use the highest possible voltage range consistent with getting a reasonable deflection, which should be **not less than** one-half of full scale.

For digital voltmeters:

- Virtually no loading effect. No difficulty in interpreting the reading, the accuracy of which then depends upon instrument design.

The ammeter

The ammeter is used to measure current **through** a component and its position in the circuit is **in series** with that component. See Figure 8.5.

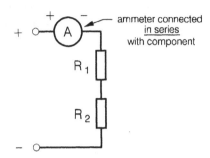

Figure 8.5 *Measurement of current*

An ammeter is a low resistance device (ideally zero) whose insertion should not affect the operation of the circuit. In practice, an ammeter will have a finite resistance and consequently there will always be a voltage drop present across it when a measurement is taken. For an analog meter this voltage can be as much as 0.5 V at full-scale deflection. The degree to which this affects the circuit operation depends upon the level of voltages in the circuit. Sometimes an alternative position for the ammeter in the circuit will overcome a possible source of error.

As examples of correct and incorrect ammeter (and voltmeter) positioning, consider the circuits shown in Figures 8.6 and 8.7. It is convenient to assume that both the ammeter and voltmeter are analog types.

Figure 8.6(a) shows the circuit of a forward biased diode (assume it to be silicon).

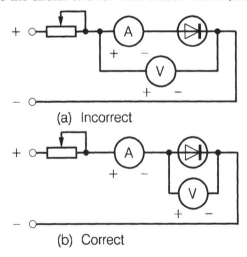

Figure 8.6 *Measurement of diode forward characteristics*

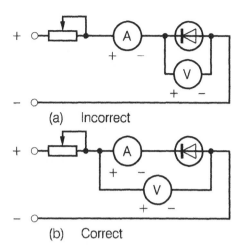

Figure 8.7 *Measurement of diode reverse characteristics*

The forward volt-drop across this diode will be 0.6 V and there will be a volt-drop of up to 0.5 V across the ammeter. The voltmeter is reading the combined voltage across the ammeter and diode in series, which would then be mistaken as the voltage across the diode itself.

The correct instrument position, shown in Figure 8.6(b), enables the voltmeter to register the diode voltage only. The voltmeter current will be small compared with the diode forward current.

Figure 8.7(a) shows the same diode in reverse bias. The reverse current through the diode will remain small as the voltage across it is increased. The ammeter is reading the diode current together with the (relatively large) voltmeter current.

The correct instrument position, shown in Figure 8.7(b), enables the ammeter to read the diode current only. The volt-drop across the ammeter will be small compared with the voltmeter reading.

The cathode ray oscilloscope

A very useful and versatile item of test equipment. It is always connected **across** the component and as with the voltmeter it is the parallel loading effect that must be considered.

The CRO has an input resistance (1 MΩ) in parallel with a capacitance of 30 pF.

The effect of a parallel resistance of 1 MΩ will follow the same pattern as that of the analog voltmeter discussed earlier.

The input capacitance will provide a low impedance path to the higher frequencies and the effects will become noticeable when the CRO displays a square or rectangular waveform.

The diagram in Figure 8.8 shows the rounded 'leading edge' which is due to a number of the high frequency components, making up the square wave, having been removed to earth by the input capacitance.

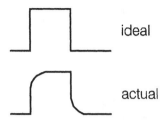

Figure 8.8 *Effect of CRO input capacitance on a square wave*

The ohmmeter

This is used to measure the resistance of so-called passive components such as resistors, capacitors and inductors. The term **passive component** is used to imply that no energy or power amplification takes place within that component. It will be appreciated that power **consumption** occurs, sometimes to the point of component burn-up.

The resistance of active components, such as transistors, can also be measured, although for this they must be in their passive or switched-off state.

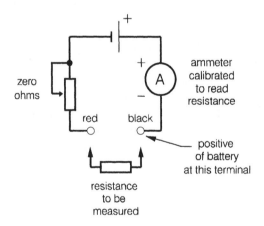

Figure 8.9 *The basic ohmmeter*

The basic circuit of an ohmmeter is shown in Figure 8.9. The necessary power supply in the form of a battery will send a current through the component whose resistance is to be measured. This current will provide a deflection of the pointer of the meter dependent on the resistance value. A variable resistor enables the pointer to be set to zero for zero external resistance.

There is usually more than one resistance range, perhaps needing batteries having several different voltages.

The scale is non-linear which makes reading accuracy difficult at the upper (left-hand) end of the scale.

Important points

For an analog meter:

- Set the pointer to read zero by connecting together the two terminals of the meter and then adjusting the 'zero ohms' control.

- Zero resistance is on the **right**-hand side of the scale (the opposite side to zero voltage and current).

- The **positive** terminal of the battery will be at the right-hand (black) terminal of the meter. This is opposite to the positive (red) terminal for voltage and current.

For a digital meter:

- Check in the operating manual for battery polarity.

Resistance measurements

A resistance measurement is a useful guide to the 'go/no-go' state of semiconductor devices.

Important points

For resistors:

- A straightforward measurement with no regard to polarity.

For capacitors:

- An ohms reading will give a measurement of insulation resistance.

- For polarized capacitors (e.g. electrolytic) observe correct polarity. Initial reading will be zero, increasing as capacitor charges up.

For diodes:

- The forward and reverse resistance should be low (Ω) and high (MΩ) respectively.
  See Figure 8.10.

For bipolar transistors:

- The forward and reverse resistances of the base–emitter and collector–base junctions should be similar to those of the diode.

- The resistance measurement between collector and emitter should be high in both directions. This measurement is useful in that it enables the base connection to be decided.

- Remember that npn and pnp transistors require an opposite polarity supply.
  See Figure 8.11.

Continued on p. 192

Important points (*Continued*)

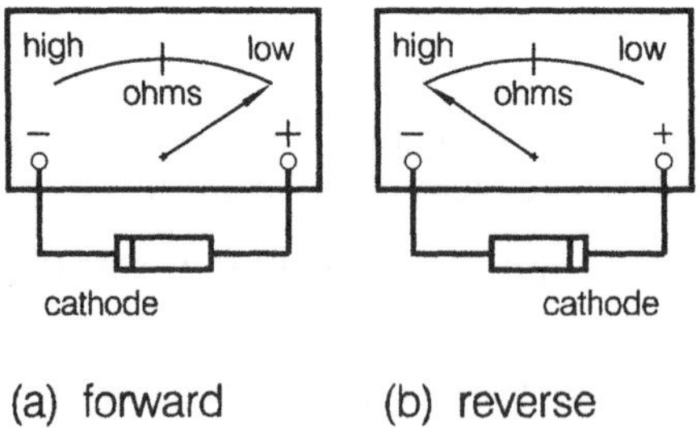

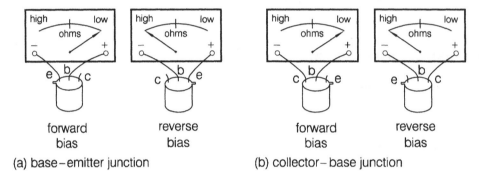

(a) forward   (b) reverse

Figure 8.10   *Testing the diode*

forward bias   reverse bias   forward bias   reverse bias

(a) base–emitter junction   (b) collector–base junction

Figure 8.11   *Testing the npn transistor*

For field effect transistors:
- The forward biased drain–gate and gate–source should be low (Ω).
- The reverse biased drain–gate and gate–source should be high (MΩ).

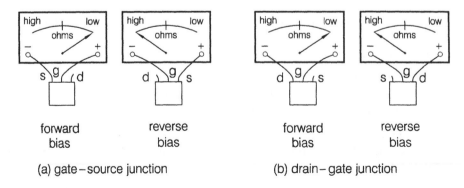

forward bias   reverse bias   forward bias   reverse bias

(a) gate–source junction   (b) drain–gate junction

Figure 8.12   *Testing the n-channel field effect transistor*

- The drain–source resistance should be low (Ω) in both directions. See Figure 8.12.

Practical Exercise 8.2

To measure the resistance of various components.
  For this exercise you will need the following components and equipment:

A collection of resistors, capacitors, inductors, diodes, and transistors (pnp and npn). (It would helpful if some components were known to be faulty.)
1 – ohmmeter

Procedure

1 Measure the DC resistance of as many components as possible and compare the values with those given in manufacturer's data.

Question

8.1 A practical exercise is undertaken in which a 22 µF capacitor is charged through a 1 MΩ resistor from a +10 V supply, as shown in Figure 8.13. The voltage across the capacitor is indicated by an analog voltmeter set to the 10 V range. The sensitivity of the meter is 20 kΩ/V and the capacitor is initially uncharged.
  The supply voltage is switched on and the capacitor starts to charge up, shown by the voltmeter reading increasing from zero. The capacitor voltage would be expected to reach a value close to +10 V after about two minutes.
  Having reached a value of +2 V quite quickly, there is now no further increase in the voltmeter reading. The supply voltage is still switched on and there is no fault in any of the components or in the circuit connections.

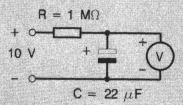

Figure 8.13   *Circuit diagram for Question 8.1*

  Suggest a likely reason for this problem and what could be done to ensure a successful outcome of the exercise.

Digital versus analog

If users were given the choice of meter, the chances are that digital would win more votes than analog. The reasoning behind this choice is probably the thought that the digital meter is bound to be easier to use and more accurate in its display. The first thought is correct, the second less so. Certainly the digital meter reading will contain no human error, and, with its constant high input resistance of 10 MΩ, will ensure minimum circuit loading on all but the highest values of resistance.

Added to this, the analog voltmeter will offer a loading resistance dependent on the range being used and the reading can sometimes be difficult to interpret.

The accuracy of the digital meter is of course determined by its internal design. However, a display of three decimal places does not, by itself, imply precision or accuracy. The sampling technique employed in providing the measurement can result in variations in one or more digits. Furthermore, when watching a meter when making adjustments, it is more difficult with a digital meter. When the actual value is less important than whether the pointer is rising or falling, the analog meter will be much appreciated.

9 Fault finding

Introduction

Fault finding, or more correctly, **logical** fault finding is the process of deducing the location and nature of the problem in a particular system. This process applies to mechanical, electrical, pneumatic, hydraulic, etc. systems and follows a universal approach.

In the general case, a system can be composed of a number of sub-systems, interconnected as necessary in order to perform the designated task. Each system or block will have a particular function, giving an expected output for a prescribed input. An example is given in Figure 9.1.

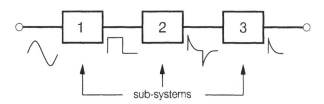

Figure 9.1 *A system*

A malfunction in **any one** block can be sufficient to prevent satisfactory operation of the system as a whole. There is hardly a need to emphasize the effect that an absence of petrol will have on motor car mobility.

Logical fault finding in complex systems must therefore consist, in order, of:

(a) identifying the faulty sub-system. This is known as fault finding at **system level**, for which a **block** diagram is required
(b) identifying the problem within that sub-system. This leads to fault finding at **component level**, for which a more detailed (schematic) diagram is needed. For electronic systems, the **circuit diagram** is an essential requirement.
(c) replacing the faulty component(s) and testing to ensure satisfactory operation. Many modern electronic circuit boards do not lend themselves very easily to component replacement, in which case, mostly for economic reasons, the remedy is to replace the entire board.

The procedures

Before any measurement testing is carried out, it can be useful to undertake a visual inspection of the unit. This is generally regarded as the **'look, smell and feel'** test.
   Look for:

(a) signs of overheating (charring/discolouring of resistors)
(b) dry joints (even in equipment which has been working)
(c) breaks in printed circuit board tracks
(d) short circuits between tracks (e.g. from solder).

   Smell for overheated components.
   Feel (with care) the surface of:

(a) transistors and ic's
(b) resistors

(for signs of excessive heat being dissipated).
   When the business of measurement finally gets under way it is important to adopt a **logical** approach, one in which every subsequent action is the result of the previous one.
   The first move (and one which is sometimes overlooked!) is to check that power is reaching the unit.
   Secondly, from **dynamic testing** with a signal generator (if appropriate) and an oscilloscope, the faulty stage should be located. The two methods for this are:

(a) tracing the signal from the **first stage** to the **last**, or vice versa. This can be long-winded especially if the fault turns out to be in the later stages. The method is useful for a small number of blocks in series.
(b) the **half-split** method. This consists of dividing the number of units in the chain into two halves, taking either one or the other, and establishing if the fault is in that particular half.
   If it is, then that half is divided into two sections and the fault condition identified as being in one or other of those two sections, and so on.
   If it is not, then the procedure is the same for the other half.
   This is the preferred method for a more complicated arrangement of blocks.

The third move is to undertake **static tests**, in order to locate the faulty component. This consists of measurements of DC voltages at appropriate points in the circuit. If values of current are required, they can usually be estimated from the measurement of voltage across a known resistance. The meter sensitivity should be at least 20 kΩ/V.

Typical fault conditions in components

(a) Resistors
   Open circuit or high in value, but very rarely short circuit

(b) Variable resistors (potentiometers)

Mechanical wear causes intermittent contact ('noisy pot') or open circuit

(c) Inductors

Open circuit, short-circuited turn (difficult to detect) or short circuit between coil and core

(d) Capacitors

Normally infinite DC resistance. Can go open circuit (difficult to detect), short circuit, or 'leaky'

(e) Diodes and transistors

Open circuit or short circuit between any two junctions.

Our objective in this chapter is to consider individual circuits and to locate the fault, with no component replacement being required. Without closer examination, measurement and sometimes component removal, the causes must remain as alternatives. It is therefore only possible, in those circumstances, to give the **most likely** reason for the fault condition.

Fault finding on components

Example 1

A number of boards have been made according to the circuit shown in Figure 9.2(a). Unfortunately each board has a single different fault.

For a correctly working board, the voltage at test point 1 with respect to the 0 V line is 11 V.

We can draw an 'equivalent circuit' as shown in Figure 9.2(b) in which the capacitors have been replaced by their equivalent DC resistance, namely infinity or open circuit.

The total resistance (R_T) can be seen from Figure 9.2(c) to be the result of 47 kΩ in series with the parallel combination of 43 kΩ and 4.7 kΩ, giving

$$R_T = 47 \text{ k}\Omega + 4.2 \text{ k}\Omega = 51.2 \text{ k}\Omega$$

The voltage (V) at test point 1 will be given by

$$V = \frac{47 \text{ k}\Omega}{51.2 \text{ k}\Omega} \times 12 \text{ V} = 11 \text{ V}$$

Let us now examine the effect on the total resistance (R_T) and the voltage (V) of two faults.

Fault 1: C_1 short circuit

The sequence of drawing the resultant equivalent circuit is shown in Figure 9.3(a) to (c).

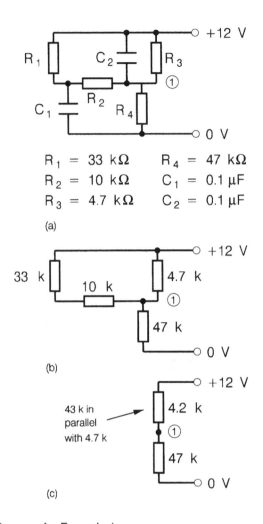

$R_1 = 33 \text{ k}\Omega$ $R_4 = 47 \text{ k}\Omega$
$R_2 = 10 \text{ k}\Omega$ $C_1 = 0.1 \text{ }\mu\text{F}$
$R_3 = 4.7 \text{ k}\Omega$ $C_2 = 0.1 \text{ }\mu\text{F}$

(a)

(b)

(c)

Figure 9.2 *Circuit diagrams for Example 1*

The total resistance R_T is 9.3 kΩ, and the voltage V is 7.6 V.

Fault 2: R_3 open circuit

The resultant circuit is shown in Figure 9.4.

It is a straightforward procedure to calculate that:

$$R_T = 90 \text{ k}\Omega \quad \text{and} \quad V = 6.3 \text{ V}$$

Notes:
- For **each** of the practical exercises that follow, there should be **only one fault at any one time**.
- The circuits relating to the questions following each practical exercise can all be made up with the component values given, in order to verify the particular fault.

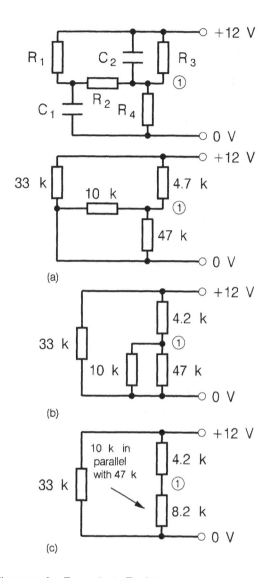

Figure 9.3 *Circuit diagrams for Example 1, Fault 1*

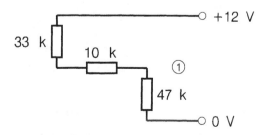

Figure 9.4 *Resultant circuit diagram for Example 1, Fault 2*

Practical Exercise 9.1

Fault finding on a component board.
For this exercise you will need the following components and equipment:

1 – resistor (4.7 kΩ, 10 kΩ, 33 kΩ, 47 kΩ)
2 – capacitor (0.1 µF)
1 – DC power supply (+12 V)
1 – DC voltmeter

Procedure

1 Make up the circuit shown in Figure 9.2(a).
2 Put on the following faults in turn, measure the voltage at test point 1 and confirm the calculations above, for:

(a) C_1 short circuit
(b) R_3 open circuit

3 Put on the following faults in turn and measure the voltage at test point 1:

(a) C_2 short circuit
(b) C_1 and C_2 short circuit (an unlikely coincidence!)

Confirm the measurements by calculation.

Questions

9.1 Refer to Figure 9.5.
   Deduce the most likely fault in each case.

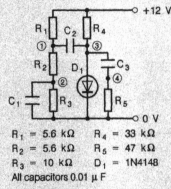

R_1 = 5.6 kΩ R_4 = 33 kΩ
R_2 = 5.6 kΩ R_5 = 47 kΩ
R_3 = 10 kΩ D_1 = 1N4148
All capacitors 0.01 µ F

(a) circuit diagram

Test point	1	2	3	4
Normal (V)	8.8	5.7	0.6	0
Fault 1 (V)	6.0	0	0.6	0
Fault 2 (V)	12.0	0	0.6	0
Fault 3 (V)	8.8	5.7	0.6	0.6
Fault 4 (V)	8.8	5.7	0	0
Fault 5 (V)	0	0	0.6	0

(b) table of values

Figure 9.5 *Circuit diagram and table of values for Question 9.1*

9.2 A box of diodes has been bought with the cathode identifying band missing. A resistance check gives the information shown in Figure 9.6. Which end is the cathode?

9.3 Ohmmeter tests on four transistors gave the information shown in Figure 9.7. State which, if any, of the transistors is faulty and give the reason(s).

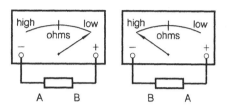

Figure 9.6 *Details for Question 9.2*

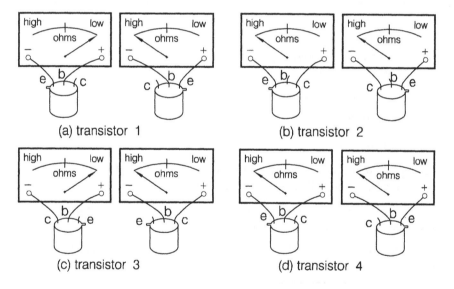

Figure 9.7 *Details for Question 9.3*

Fault finding on DC power supplies

Practical Exercise 9.2

Fault finding on a bridge rectifier power supply.
For this exercise you will need the following components and equipment:

Continued on p. 202

Practical Exercise 9.2 *(Continued)*

1 – transformer unit (240 V primary, 12 V secondary with centre-tap)
4 – silicon diode (1N4001)
1 – filament lamp (6 V)
1 – capacitor (1000 μF, 16 V)
1 – double beam cathode ray oscilloscope
1 – DC voltmeter

Procedure

1 Make up the circuit shown in Figure 2.20.
2 Measure the DC output voltage and the peak-to-peak ripple voltage, which should be approximately 8 V and 0.5 V respectively. Regard these as 'normal' readings.
3 Put on the following faults in turn, measure the above quantities and note the output waveform, for:
   (a) C_1 open circuit (remove C_1)
   (b) D_1 open circuit (remove D_1)
   (c) D_1 short circuit.

Important points

You should find that:

- An open-circuit reservoir capacitor will give a low DC output voltage resulting from the full-wave 'pulsating DC'.

- An open-circuit diode will result in a half-wave output, giving a slightly reduced DC voltage but an increase in ripple.

- A short-circuit diode will give virtually the same result as one on open circuit.

Questions

Questions 9.4 and 9.5 refer to Figure 2.20.

9.4 Why does a short-circuit diode give a half-wave rather than a full-wave output?
   Hint. Redraw the bridge circuit with, say, D_3 replaced by a link.
9.5 Describe the effect on the circuit operation, of diodes D_3 and D_4 being short circuit **at the same time**.

Questions 9.6 and 9.7 refer to Figure 2.24.

Practical Exercise 9.3

Fault finding on a series voltage regulator.
For this exercise you will need the following components and equipment:

1 – zener diode (4.7 V, 500 mW)
1 – transistor (2N3053)
1 – transistor (BC109)
1 – resistor (470 Ω, 1 W)
1 – resistor (1 kΩ, 0.5 W)
2 – resistor (10 kΩ, 0.5 W)
1 – DC power supply (variable from +15 V to +25 V)
1 – DC voltmeter

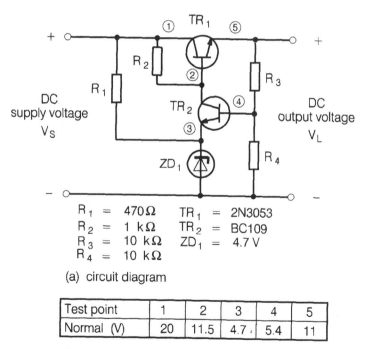

R_1	=	470 Ω	TR_1	=	2N3053
R_2	=	1 kΩ	TR_2	=	BC109
R_3	=	10 kΩ	ZD_1	=	4.7 V
R_4	=	10 kΩ			

(a) circuit diagram

Test point	1	2	3	4	5
Normal (V)	20	11.5	4.7	5.4	11

(b) table for normal no-load operation

Figure 9.8 *Circuit diagram and table of values for Practical Exercise 9.3*

Continued on p. 204

Practical Exercise 9.3 *(Continued)*

Procedure

1 Make up the circuit shown in Figure 9.8(a). The pin connection diagram for the 2N3053 transistor is given in Figure 2.37 and for the BC109 transistor in Figure 2.40.
2 Set the DC supply voltage (V_S) to +20 V, and take voltage readings at the five test points. These may be regarded as 'normal readings' and should be similar to those given in the table in Figure 9.8(b).

Important points

- The voltage at the junction of R_3 and R_4 will be one-half of the output voltage (V_L), assuming R_3 and R_4 are equal. This voltage must be more positive than the constant voltage at the base of TR_2 in order for TR_2 to conduct.

- The base-emitter voltage (V_{be2}) is the input signal of the **amplifier** transistor TR_2 which has resistor R_2 as its collector load.

- The collector voltage (V_{c2}) of TR_1 is the amplifier output of the input voltage V_{be2}.

- The series transistor (TR_1) behaves as an emitter follower, with the voltage at its emitter (which equals V_L) always being just less (typically 0.6 V) than the voltage at its base.

- If V_L **increases**, there will be an **increase** in V_{be2} causing a **decrease** (remember 180° phase shift) in V_{c2} and in the base voltage of TR_1.

- The resultant decrease in V_L should, for a perfect regulator, be equal to the original increase.

---

3 Put on the following faults in turn and measure the test point voltages, for:
(a) ZD_1 short circuit
(b) R_4 open circuit
(c) TR_2 collector–emitter short circuit

Important points

- A reference voltage of zero at test point 2 indicates the presence of a short circuit. An **open** circuit at that point would provide a substantial test point voltage following completion of the circuit by the meter.

- Following an open circuit in R_4, the base voltage of TR_2 will initially rise, causing TR_2 to conduct heavily (V_{ce2} falls). The output voltage V_L will then fall.

- When two points are at identical voltages (apart from zero) there are strong indications of a short circuit. Measurements at test points 2 and 3 illustrate this.

- In general, a number of test points being at zero volts is suggestive of an open circuit.

Question

9.8 Test point voltages for five separate faults for the circuit of Figure 9.8(a) are given in the table shown in Figure 9.9.
Deduce the most likely fault in each case.

Test point	1	2	3	4	5
Normal (V)	20	11.5	4.7	5.4	11
Fault 1 (V)	20	20	4.7	0	19.5
Fault 2 (V)	20	0	4.7	0	0
Fault 3 (V)	20	10.6	4.3	4.9	10
Fault 4 (V)	20	20	19.9	9.6	19.4
Fault 5 (V)	20	20	4.7	0	0

Figure 9.9 *Table of values for Question 9.8*

Fault finding on transistor amplifiers

The reader will recall that the first requirement of any amplifier is that its DC biasing conditions must be correctly set up for the type of operation expected.

These bias conditions would normally be provided by way of test point voltages, but in their absence, they can be estimated fairly easily. This was shown in Chapter 3. In most cases the DC voltage levels will lead to the particular fault, although the waveform of the output signal can occasionally provide the only clue.

Practical Exercise 9.4

Fault finding on a common emitter transistor amplifier.
For this exercise you will need the following components and equipment:

1 – npn transistor (BC109)
1 – resistor (470 Ω, 1.5 kΩ, 10 kΩ, 47 kΩ)
3 – capacitor (100 μF, 16 V)
1 – DC power supply (+12 V)
1 – audio frequency signal generator
1 – double beam cathode ray oscilloscope
1 – DC voltmeter

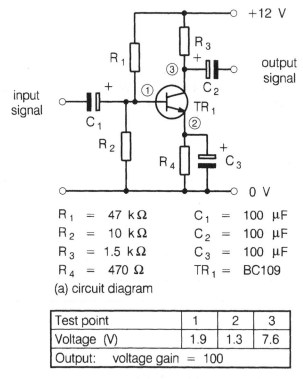

R_1 = 47 kΩ C_1 = 100 μF
R_2 = 10 kΩ C_2 = 100 μF
R_3 = 1.5 kΩ C_3 = 100 μF
R_4 = 470 Ω TR_1 = BC109

(a) circuit diagram

Test point	1	2	3
Voltage (V)	1.9	1.3	7.6
Output: voltage gain = 100			

(b) table for normal operation

Figure 9.10 *Circuit diagram and table of values for Practical Exercise 9.4*

Procedure

1 Make up the circuit shown in Figure 9.10(a). The pin connection diagram for the BC109 transistor is given in Figure 2.40.

2 Measure the DC voltages at the three test points, and the voltage gain of the amplifier at 1 kHz. Typical values are given in Figure 9.10(b).

3 Put on the following faults in turn, measure the test point voltages and note the output waveform, for:

(a) R_1 open circuit
(b) C_2 open circuit (remove C_2)
(c) TR_1 base–emitter short circuit (connect a link between these two electrodes).

Important points

You should find that:

- All of the above faults result in **no signal output**. By itself, that symptom is only partly helpful since the fault(s) could be due either to an incorrect DC condition (**DC fault**) or to an open circuit in the signal path (**AC fault**).

- An open-circuit upper bias resistor will leave the transistor unbiased, with both base and emitter being at zero volts.
  The transistor will be cut off with its collector being at $+V_{CC}$.

- All test points at their 'normal' value indicates an AC fault.
  In the case of an open-circuit coupling capacitor, tracing the signal along its series path will find the culprit.
  Decoupling capacitors offer a bypass route for a fraction of the signal voltage, which would otherwise provide a measure of negative feedback giving reduced amplifier gain.

- Two test points at identical voltages (not zero) are a good sign of a short circuit.
  A transistor base–emitter short circuit will cause the base voltage to fall drastically, since the lower base bias resistor and the emitter resistor are now effectively in parallel. The transistor will be cut off.
  For this example, the parallel equivalent of R_2 and R_4 is 0.45 kΩ, with the resultant voltage at test point 1 calculated as

$$\frac{0.45 \text{ k}\Omega}{47 \text{ k}\Omega} \times 12 \text{ V}, \text{ or } 0.11 \text{ V}.$$

Questions

9.9 Test point voltages for seven separate faults for the circuit of Figure 9.10(a) are given in the table shown in Figure 9.11.
Deduce the most likely fault in each case.

Continued on p. 208

Questions *(Continued)*

	Test point	1	2	3
Normal	Voltage (V)	1.9	1.3	7.6
	Output: voltage gain = 100			
Fault 1	Voltage (V)	0.7	0.2	0.1
	Output: none			
Fault 2	Voltage (V)	1.9	1.5	12
	Output: none			
Fault 3	Voltage (V)	1.9	1.3	7.6
	Output: voltage gain = 3			
Fault 4	Voltage (V)	3.3	2.6	3.3
	Output: none			
Fault 5	Voltage (V)	0.7	0	0.1
	Output: very low voltage gain			
Fault 6	Voltage (V)	3.6	2.9	3.0
	Output: very low voltage gain			
Fault 7	Voltage (V)	2.1	2.8	2.8
	Output: none			

Figure 9.11 *Table of values for Question 9.9*

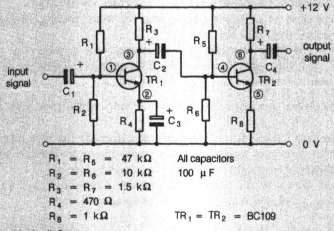

$R_1 = R_5 = 47\ k\Omega$ All capacitors
$R_2 = R_6 = 10\ k\Omega$ 100 μF
$R_3 = R_7 = 1.5\ k\Omega$
$R_4 = 470\ \Omega$
$R_8 = 1\ k\Omega$ $TR_1 = TR_2 = BC109$

(a) circuit diagram

	Test point	1	2	3	4	5	6
Normal	Voltage (V)	1.9	1.3	7.6	1.9	1.3	10.1
	Output: voltage gain = 150						
Fault	Voltage (V)	1.9	1.3	5.9	5.9	5.2	4.5
	Output: low voltage gain (approx 30)						

(b) table of values

Figure 9.12 *Circuit diagram and table of values for Question 9.10*

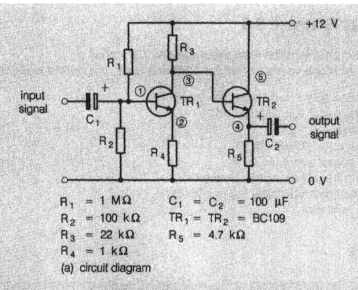

R₁ = 1 MΩ   C₁ = C₂ = 100 μF
R₂ = 100 kΩ   TR₁ = TR₂ = BC109
R₃ = 22 kΩ   R₅ = 4.7 kΩ
R₄ = 1 kΩ
(a) circuit diagram

	Test point	1	2	3	4
Normal	Voltage (V)	0.8	0.3	3.4	3.0
	Output:	voltage gain = 20			
Fault 1	Voltage (V)	0	0	11.5	11.0
	Output:	none			
Fault 2	Voltage (V)	0.01	0.01	11.5	11.0
	Output:	none			
Fault 3	Voltage (V)	0.5	0.02	0.01	0
	Output:	none			
Fault 4	Voltage (V)	0.8	0.4	3.4	3.2
	Output:	none			
Fault 5	Voltage (V)	0.8	0.6	11.5	11.2
	Output:	none			

(b) table of values

Figure 9.13 *Circuit diagram and table of values for Question 9.11*

9.10 A faulty two-stage amplifier has the circuit diagram and test point voltages given in Figure 9.12.

(a) Deduce the fault.

(b) Calculate, for the 'normal' amplifier, the anticipated voltage gain of each stage.

Hint. Start from the second amplifying stage.

9.11 A number of faulty two-stage amplifiers with direct coupling between stages have the circuit diagram and test point voltages given in Figure 9.13. For each amplifier deduce the most likely fault.

Practical Exercise 9.5

Fault finding on a junction field effect transistor amplifier.
For this exercise you will need the following components and equipment:

1 – junction field effect transistor (2N3819)
1 – resistor (4.7 kΩ, 33 kΩ, 1 MΩ)
2 – capacitor (0.1 μF)
1 – capacitor (100 μF, 25 V)
1 – DC power supply (+15 V)
1 – audio frequency signal generator
1 – double beam cathode ray oscilloscope
1 – DC voltmeter

Procedure

1 Make up the circuit shown in Figure 9.14(a). The pin connection diagram for the
2N3819 transistor is given in Figure 3.32.
2 Measure the DC voltages at the two test points, and the voltage gain of the
amplifier at 1 kHz. Typical values are given in Figure 9.14(b).

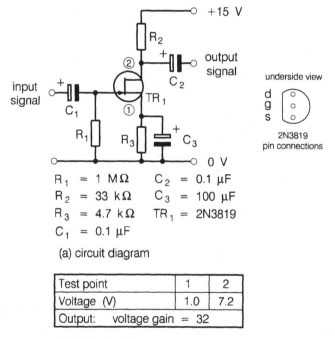

R_1 = 1 MΩ C_2 = 0.1 μF
R_2 = 33 kΩ C_3 = 100 μF
R_3 = 4.7 kΩ TR_1 = 2N3819
C_1 = 0.1 μF

(a) circuit diagram

Test point	1	2
Voltage (V)	1.0	7.2
Output: voltage gain = 32		

(b) table of values

Figure 9.14   *Circuit diagram and table of values for Practical Exercise 9.5*

Important points

- The gate–source bias voltage will be measured at test point 1, as the DC voltage across the source resistor R_3. The gate is **negative** with respect to the source and any decrease in this voltage will result in an increase in drain current.
- A test point at the drain would be unhelpful, since there will be no voltage across resistor R_1 (1 MΩ), which is a DC return for the gate–source bias voltage.
- The AC faults involving open-circuit capacitors will provide similar symptoms to the bipolar transistor amplifier.

3 Put on the following faults in turn, measure the test point voltages and note the output waveform, for:

(a) R_2 open circuit
(b) R_3 open circuit.

Important points

- Zero volts at both drain and source suggests an open-circuit feed from the $+V_{DD}$ supply. Fault (a).

- An open-circuit source resistor will result in zero drain current, the transistor being cut off and its drain voltage rising to $+V_{DD}$. The meter connection causes an increase in source voltage, and the drain voltage to be just less than cut-off value. Fault (b).

Question

	Test point	1	2
Normal	Voltage (V)	1.0	7.2
	Output: voltage gain = 32		
Fault 1	Voltage (V)	1.0	7.2
	Output: voltage gain = 5		
Fault 2	Voltage (V)	1.8	2.0
	Output: none		
Fault 3	Voltage (V)	0	0.1
	Output: none		
Fault 4	Voltage (V)	1.0	7.2
	Output: none (gain = 32)		

Figure 9.15 *Table of values for Question 9.12*

Continued on p. 212

Question (*Continued*)

9.12 Test point voltages for four separate faults for the circuit of Figure 9.14(a) are given in the table shown in Figure 9.15.
Deduce the most likely fault in each case.

Fault finding on operational amplifiers

The fault finding technique for equipment using integrated circuits is very similar to that using discrete components. The initial checks, together with the dynamic and static measurements, are all valid weapons in the task of locating the troublesome area.

The main difference concerns the integrated circuit itself, which although physically robust and generally reliable in performance, needs to be treated with care.

Unlike its digital counterpart, the analog or linear ic will operate over a wider range of supply voltages.

The ic contains both active and passive components which together form a functioning block. Any fault within the ic itself means that it must be replaced.

Important points

- Handle by the case and avoid touching the pins wherever possible.
- Ensure that pins do not get bent over and that they are all connected to the pcb or ic holder.
- Check that power is reaching the pins of the ic.
- When taking measurements, be careful not to short together adjacent pins.
- Avoid excessive heat if unsoldering the ic.

Practical Exercise 9.6

Fault finding on an inverting AC operational amplifier.
For this exercise you will need the following components and equipment:

1 – ic operational amplifier (741)
2 – resistor (1 kΩ)
1 – resistor (4.7 kΩ)
2 – capacitor (10 µF, 16 V)
1 – DC power supply (±15 V)
1 – audio frequency signal generator
1 – double beam cathode ray oscilloscope
1 – DC voltmeter

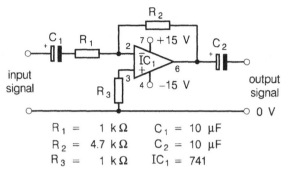

$$R_1 = 1 \text{ k}\Omega \qquad C_1 = 10 \text{ μF}$$
$$R_2 = 4.7 \text{ k}\Omega \qquad C_2 = 10 \text{ μF}$$
$$R_3 = 1 \text{ k}\Omega \qquad IC_1 = 741$$

(a) circuit diagram

	Pin number				Input signal	Output signal
	7 (V)	4 (V)	2 (V)	3 (V)	6 (V)	(V)	(V)
Normal	+15	−15	0**	0	0	1*	4.7*

** virtual earth   * peak-to-peak

(b) table of values

Figure 9.16 *Circuit diagram and table of values for Practical Exercise 9.6*

Procedure

1 Make up the circuit shown in Figure 9.16(a). The pin connections for the 741 ic are shown in Figure 6.5.
2 Measure the DC voltages at the two supply pins of the ic (7 and 4 respectively) and the voltage gain of the amplifier at 1 kHz. Typical values are given in Figure 9.16(b).
3 Put on the following faults in turn, and take measurements to form a similar table to that shown in Figure 9.16(b), for:

(a) R_2 open circuit
(b) R_3 open circuit
(c) pin 2 open circuit (disconnect pin 2 from the junction of R_1 and R_2)
(d) a faulty ic (remove it from the circuit!).

Important points

- An open-circuit feedback resistor provides a theoretical amplifier gain of infinity. In practice the output signal will always be limited (saturated) to approximately 2 V below $\pm V_S$.

Continued on p. 214

Practical Exercise 9.6 (*Continued*)

Important points (*Continued*)

- An open circuit to the non-inverting input will leave this input 'floating'. The amplifying function ceases and the DC level at the output will fall to approximately $-V_S$ with the inverting input following, through R_2.

- An open circuit to the inverting input (now floating) will cause the DC level at the output and at the junction of R_1/R_2 to rise to approximately $+V_S$.
  The input signal will feed forward through R_2, giving an identical output signal. This latter symptom will also result from an open circuit within the ic.

Questions

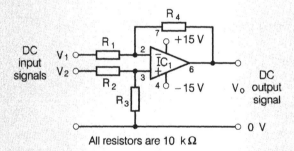

All resistors are 10 kΩ

(a) circuit diagram

| | Pin number | | | | Inputs | | Output | Additional |
	7 (V)	4 (V)	2 (V)	3 (V)	V_1	V_2	V_0 (V)	measurement (V)
Normal	+15	−15	+2.5	+2.5	+5	+5	0	
Fault 1	+15	−15	+2.5	+2.5	+5	+5	+2.5	
Fault 2	+15	−15	0	0	+5	+5	+5	
Fault 3	+15	−15	−0.3	+2.5	+5	+5	+13	junction of R_1/R_4 +9.6
Fault 4	+15	−15	+5	+2.5	+5	+5	−13	
Fault 5	+15	−15	−4	−0.1	+5	+5	−13	junction of R_2/R_3 +2.5

(b) table of values

Figure 9.17   *Circuit diagram and table of values for Question 9.13*

9.13 Five differential amplifiers, built to the circuit shown in Figure 9.17(a), each have a single different fault and the symptoms are given in Figure 9.17(b). Deduce the most likely fault in each case.

9.14 Fault conditions for two comparator amplifiers having the circuit shown in Figure 9.18(a) are given in Figure 9.18(b). Deduce the fault for each.

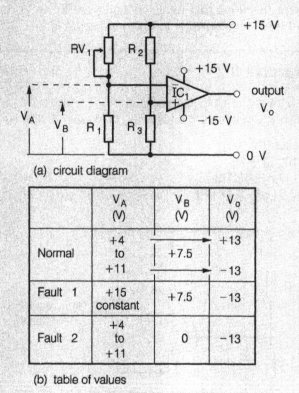

(a) circuit diagram

	V_A (V)	V_B (V)	V_o (V)
Normal	+4 to +11	+7.5	+13 −13
Fault 1	+15 constant	+7.5	−13
Fault 2	+4 to +11	0	−13

(b) table of values

Figure 9.18   *Circuit diagram and table of values for Question 9.14*

Fault finding on oscillators

Important points

The requirements are:

- an **amplifying** section and a **feedback** section
- a specified amount of **positive feedback** from output to input.

The requirement is met by:

- the amplifying section having sufficient, but not too much, **gain**

Continued on p. 216

Important points (*Continued*)

- the feedback section having the **phase shift** necessary to give an overall phase shift of 0°.

It should be apparent therefore, that oscillator faults are likely to concentrate on one or other, or both, of these requirements not having been met.

Practical Exercise 9.7

Fault finding on a transistor Colpitts oscillator.
For this exercise you will need the following components and equipment:

1 − npn transistor (BC109)
1 − inductor (100 μH)
1 − capacitor (47 nF, 100 nF)
1 − capacitor 0.01 μF, 0.47 μF)

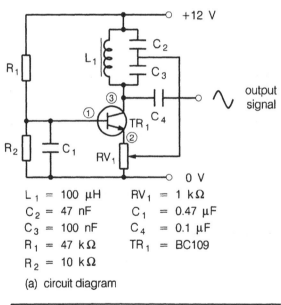

$$L_1 = 100\ \mu H \qquad RV_1 = 1\ k\Omega$$
$$C_2 = 47\ nF \qquad C_1 = 0.47\ \mu F$$
$$C_3 = 100\ nF \qquad C_4 = 0.1\ \mu F$$
$$R_1 = 47\ k\Omega \qquad TR_1 = BC109$$
$$R_2 = 10\ k\Omega$$

(a) circuit diagram

	Test point	1	2	3
Normal	Voltage (V)	1.9	1.3	12
	Output: 6 V pk−pk at 95 kHz			

(b) table of values

Figure 9.19   *Circuit diagram and table of values for Practical Exercise 9.7*

1 – resistor (10 kΩ, 47 kΩ)
1 – variable resistor (1 kΩ)
1 – DC power supply (+12 V)
1 – double beam cathode ray oscilloscope
1 – DC voltmeter

Procedure

1 Connect up the circuit shown in Figure 9.19(a). The pin connection diagram for the BC109 transistor is shown in Figure 2.40.
2 Adjust RV_1 so that the unit oscillates. Measure the frequency of oscillation.
3 Return to the non-oscillating state and measure the DC voltages at the three test points.
   Typical 'normal' values are given in Figure 9.19(b).

Important points

- When the unit oscillates, a voltage measurement at test point 3 will not represent the true DC condition.

- The low DC resistance of the coil (L) means that the collector voltage will be at $+V_{CC}$.

4 Put on the following faults in turn, measure the test point voltages and note the output waveform, for:

(a) L_1 open circuit (release one end or the other)
(b) C_2 (or C_3) short circuit.

Important points

- With an open-circuit collector load, any voltage present at the collector will be due to the meter connection.

- A short-circuited tuning capacitor will feed the $+V_{CC}$ supply to the emitter through RV_1, whose slider position will determine the reading at test point 2.

Question

9.15 Details of three separate faults for the circuit of Figure 9.19(a) are given in the table shown in Figure 9.20.

Continued on p. 218

Question (*Continued*)

Deduce the most likely fault in each case.

	Test point	1	2	3
Normal	Voltage (V)	1.9	1.3	12
	Output: 6 V pk−pk at 95 kHz			
Fault 1	Voltage (V)	0	0	12
	Output: none			
Fault 2	Voltage (V)	0.3	0.3	12
	Output: none			
Fault 3	Voltage (V)	1.9	1.3	12
	Output: none			

Figure 9.20   *Table of values for Question 9.15*

Important point

- An oscillator, which is an example of a closed-loop system, can be tested by:
  (a) breaking the loop at any point,
  (b) injecting the appropriate signal at the input side of the break, and,
  (c) measuring the output signal at the other side of the break.

The Wien Bridge oscillator can be tested in this way, as the following practical exercise will demonstrate.

Remember that the requirements in this case, for oscillation at a specific frequency, are:

(a) a minimum amplifier gain of three, with zero phase shift
(b) a feedback network with an attenuation of three (or gain = one-third) and zero phase shift

and resulting from (a) + (b)

(c) an overall gain of not less than 1, with zero phase shift.

Practical Exercise 9.8

Fault finding on a integrated circuit Wien Bridge oscillator.
  For this exercise you will need the following components and equipment:

1 − ic operational amplifier (741)

2 – resistor (4.7 kΩ)
1 – resistor (1 kΩ, 1.8 kΩ)
1 – variable resistor (4.7 kΩ)
2 – capacitor (0.01 μF)
1 – capacitor (100 μF, 25 V)
2 – diode (1N4148)
1 – DC power supply (±15 V)
1 – audio frequency signal generator
1 – double beam cathode ray oscilloscope
1 – DC voltmeter

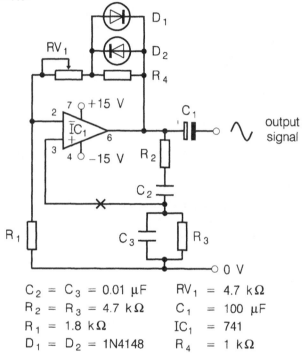

$$C_2 = C_3 = 0.01 \ \mu F \qquad RV_1 = 4.7 \ k\Omega$$
$$R_2 = R_3 = 4.7 \ k\Omega \qquad C_1 = 100 \ \mu F$$
$$R_1 = 1.8 \ k\Omega \qquad IC_1 = 741$$
$$D_1 = D_2 = 1N4148 \qquad R_4 = 1 \ k\Omega$$

Figure 9.21   *Circuit diagram for Practical Exercise 9.8*

Procedure

1 Connect up the circuit shown in Figure 9.21. The pin connection diagram for the 741 ic is shown in Figure 6.5.
2 If the unit is not already oscillating, vary RV_1 to give the maximum undistorted sinusoidal output. Use the CRO to measure the frequency of oscillation and the voltage output.

To test the individual sections of this oscillator:

3 Break the loop at point X.

Continued on p. 220

Practical Exercise 9.8 (*Continued*)

4 Connect the signal generator to the non-inverting input side of the break, together with one input of the CRO. For these measurements, the frequency of the input signal should be varied around the theoretical frequency of oscillation.

5 With RV_1 set to a maximum value, connect the second input of the CRO to the output of the amplifier and estimate the voltage gain.

In practice, this will need to be a little larger than three. Check that this is possible.

6 Transfer the second input of the CRO to the junction of C_2/C_3 (the other side of the break). Estimate the frequency at which the signal at the junction is in phase with that from the output of the amplifier. Check that the attenuation of this $2RC$ network is then three.

The frequency here should be close in value to that measured in Procedure 2. In practice the actual frequency is likely to be slightly higher than the theoretical value, due mainly to resistance loading effects.

7 Remove the signal generator and make good the circuit.

Clearly, if the separate requirements of (a) and (b) above are not met, the individual sections can be examined in more detail.

Aside from this, several faults can now be put on your (working) oscillator.

8 Put on the following faults in turn, note the output waveform and measure the DC voltage at pin 6 of the op-amp, for:

(a) R_1 open circuit – measure pin 6 of op-amp
(b) pin 2 open circuit – measure pin 6
(c) pin 3 open circuit – measure pin 6.

Important points

- With an open-circuit feedback resistor, the amplifier output will be saturated and will bear little relation to the correct waveform. The DC voltage at pin 6 will represent the average value of this waveform.

- An open-circuit inverting input will cause the output to go to (approximately) $+V_S$.

- An open-circuit non-inverting input will cause the output to go to (approximately) $-V_S$.

Question

Refer to Figure 9.21.

9.16 What will be the likely effect on the output waveform of the following separate faults:

Practical Exercise 9.9

Fault finding on a transistor astable multivibrator.
For this exercise you will need the following components and equipment:

2 – npn transistor (BC109)
2 – resistor (1 kΩ)
2 – resistor (33 kΩ)
2 – capacitor (0.01 µF)
1 – DC power supply (+12 V)
1 – double beam cathode ray oscilloscope
1 – DC voltmeter

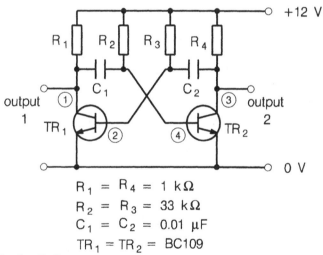

$$R_1 = R_4 = 1 \text{ k}\Omega$$
$$R_2 = R_3 = 33 \text{ k}\Omega$$
$$C_1 = C_2 = 0.01 \text{ µF}$$
$$TR_1 = TR_2 = BC109$$

(a) circuit diagram

	Test point	1	2	3	4
Normal	Voltage	+5.6	−1.5	+5.6	−1.5
	Output:	square, 12 V pk–pk, 2275 Hz			

(b) table of values

Figure 9.22 *Circuit diagram and table of values for Practical Exercise 9.9*

Continued on p. 222

Practical Exercise 9.9 (*Continued*)

Procedure

1 Make up the circuit of Figure 9.22(a). The pin connection diagram for the BC109 transistor is shown in Figure 2.40.
2 Display the waveform from **output 1** on the CRO.
   Measure the frequency of this waveform and the DC voltages at the four test points.
   These DC voltages represent the average DC level at the respective points. Typical values are given in Figure 9.22(b).
3 Put on the following faults in turn, measure the test point voltages and note the output waveform from **output 1,** for:

(a) TR_1 base–emitter open circuit (open the emitter connection to the 0 V line)
(b) TR_2 base–emitter short circuit
(c) R_2 open circuit.

Important points

- An open-circuit base–emitter junction will cause the base of that transistor to rise to almost $+V_{CC}$, with the transistor therefore being in the 'off' state. The other transistor will be taken to the 'on' state through its bias resistor. No output waveform.

   A voltage measurement at the emitter of the faulty transistor would reveal a value 0.6 V below that of its base. This confirms the method of providing the fault, namely, open-circuiting the **external** emitter connection. A bona fide **internal** base–emitter open circuit would give an emitter voltage of zero.

- A short-circuit base–emitter junction will keep that transistor turned off, while the other transistor will be turned on through its bias resistor.

   The alternative fault of an open-circuit bias resistor to this **same** transistor will give identical test point readings.

   An in-circuit resistance measurement would decide.

   No output waveform.

Question

9.17 Details of four faults for the astable multivibrator of Figure 9.22(a) are given in Figure 9.23.
   Deduce the nature of each fault.

	Test point	1	2	3	4
Normal	Voltage	+5.6	−1.5	+5.6	−1.5
	Output: square, 12 V pk−pk, 2275 Hz				
Fault 1	Voltage	+0.1	+0.7	+0.1	+0.7
	Output: none				
Fault 2	Voltage	+12	+0.1	+0.1	+0.7
	Output: none				
Fault 3	Voltage	+9.0	−2.9	+2.7	−0.2
	Output: rect, 12 V, m:s 3.3:1, 1190 Hz				
Fault 4	Voltage	+12	+11.5	+0.1	+0.7
	Output: none				

Figure 9.23 *Table of values for Question 9.17*

Answers to questions

Chapter 2

2.2 (a) L_1
 (b) both L_1 and L_2
 (c) L_1
2.3 (d) D_1 and D_4
2.4 (e) L_1 6 V L_2 6 V
 (f) L_1 0.6 V, L_2 11.3 V
 (g) L_1 0.6 V, L_2 11.3 V
2.6 23.4 V. Assumptions:

   (i) capacitor charges up to the peak of the supply minus the volt-drop across forward-conducting diode.
   (ii) diode volt-drop $= 0.6$ V.

   Minimum PIV rating 48 V
2.7 17.8 V rms, 25.2 V
2.13 16.7%
2.14 50%
2.15 30 Ω, 0.3 W
2.16 (a) 500 Ω
 (b) ± 40 mV

Chapter 3

3.1 2.5 mA
3.2 5 μA
3.3 125
3.4 (c) 15 mA
 (d) 100
 (e) (18.6 mA), 93
 (f) (18.6 V), 23.25
 (g) 2162.25
3.5 (b) 8.2 mA, 6.4 V
 (c) 31.5
 (e) 160 μA
3.6 12

3.7 (a) 1.5 kΩ
   (b) 80.4
   (c) −252
3.8 (a) 4.8 kΩ
   (b) 66.7 kΩ
   (c) 187.2
   (d) −183.3
3.9 (a) 63.9
   (b) 11.1
3.10 (a) 22.7
   (b) 25

Chapter 4

4.1 (a) 30 dB
   (b) 27 dB
   (c) 24 dB
   (d) 0 dB
   (e) −10 dB
   (f) −20 dB
4.2   100 W
4.3   38 dB, 6.31 W
4.4   0.67 W, 21.8 dB
4.5   Yes, since specification gives 10 W at 100 Hz and 18 kHz
4.6   25.1 mV

Chapter 5

5.1   80
5.2   1/20
5.3   800
5.4   49.88, 49.75
5.5   1/50
5.9   Amp 1. $A = 940,\ \ A' = 19.58$
      Amp 2. $A = 1880,\ A' = 19.79$
      Amp 3. $A = 2820,\ A' = 19.86$
      Amp 4. $A = 3760,\ A' = 19.89$
      300%, 1.58%
5.10 20.06 dB.
      ($\beta = 0.099$, A goes from 1000 to 3162, A' from 10 to 10.07)

Chapter 6

6.1 (a) −1 V
   (b) +2 V

6.2 $10 \text{ k}\Omega$
6.3 $R_2 = 1 \text{ k}\Omega$ fixed $+10 \text{ k}\Omega$ variable
6.4 $3.1, -0.62$ V
6.5   If $R_2 = 1 \text{ k}\Omega$, then $R_3 = 47 \text{ k}\Omega$
6.6   If $R_3 = 47 \text{ k}\Omega$, gain variation is from 1 to 11
6.7 -0.66 V
6.8 $+1.28$ V
6.9   See Figure A.1.

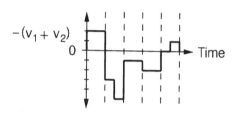

Figure A.1   *Chapter 6, Question 6.9*

6.10 (a) If $R_4 = 10 \text{ k}\Omega$, then

for input 1, $R_1 = 20 \text{ k}\Omega$

for input 2, $R_1 = 4 \text{ k}\Omega$

for input 3, $R_1 = 2 \text{ k}\Omega$

(b) $+0.5$ V
6.11 $+235$ mV

Chapter 7

7.4 (a) 13.8 nF
(b) 276 Ω
7.5 $10.6 \text{ k}\Omega$
7.6   1591 kHz, 1:100
7.7   396 pF

Chapter 9

9.1 (1) C_1 short circuit (unlikely to be R_3)
(2) R_2 open circuit
(3) C_3 short circuit
(4) D_1 short circuit
(5) R_1 open circuit (unlikely to be $C_3 + R_5$)
9.2 A

9.3   Transistor 4 is faulty. The forward biased base–emitter junction should be low resistance.

9.8 (1) R_3 open circuit
   (2) R_2 open circuit
   (3) R_1 open circuit
   (4) ZD_1 open circuit
   (5) TR_1 base–emitter open circuit

9.9 (1) R_3 open circuit
   (2) R_4 open circuit
   (3) C_3 open circuit
   (4) TR_1 collector–base short circuit
   (5) C_3 short circuit
   (6) R_2 open circuit
   (7) TR_1 collector–emitter short circuit

9.10 (a) C_2 short circuit
   (b) Gain of second stage $= R_7/R_8$

9.11 (1) R_1 open circuit
   (2) TR_1 base–emitter short circuit
   (3) R_3 open circuit
   (4) R_5 open circuit
   (5) R_4 open circuit

9.12 (1) C_3 open circuit
   (2) TR_1 gate–source short circuit
   (3) C_3 short circuit
   (4) C_2 open circuit

9.13 (1) R_1 open circuit
   (2) R_2 open circuit
   (3) pin 2 open circuit
   (4) R_4 open circuit
   (5) pin 3 open circuit

9.14 (1) R_1 open circuit
   (2) R_2 open circuit

9.15 (1) R_1 open circuit
   (2) TR_1 base–emitter short circuit
   (3) C_2 or C_3 or connection to RV_1 slider, open circuit

9.16 (a) Negative half-cycle saturates before positive half-cycle
   (b) Reduction in gain. Recovered by adjustment of RV_1
   (c) Output waveform distorted in 'crossover' region

9.17 (1) C_1 open circuit
   (2) C_2 short circuit
   (3) R_3 high resistance (100 kΩ)
   (4) R_1 open circuit

Index

Printed and bound by CPI Group (UK) Ltd, Croydon, CR0 4YY

17/10/2024

01775697-0009